S0-CBW-741

Advances in

ECOLOGICAL RESEARCH

VOLUME 1

Advances in ECOLOGICAL RESEARCH

Edited by

J. B. CRAGG

The Nature Conservancy, Merlewood Research Station, Grange-over-Sands, Lancashire, England

VOLUME 1

1962

ACADEMIC PRESS
London and New York

ACADEMIC PRESS INC. (LONDON) LTD.
Berkeley Square House
London, W.1

U.S. Edition, Published by
ACADEMIC PRESS INC.
111 Fifth Avenue New York 3, New York

Library of Congress Catalog Card Number: 62–21479

Printed in Great Britain by Robert MacLehose & Co. Ltd, Glasgow, Scotland

Contributors to Volume 1

J. D. OVINGTON, *The Nature Conservancy, Monk's Wood Research Station, St. Ives, Huntingdonshire, England.*

A. MACFADYEN, *Department of Zoology, University College, Swansea, England.*

M. E. D. POORE, *Botany Department, University of Malaya, Kuala Lumpur, Malaya.*

L. B. SLOBODKIN, *Department of Zoology, University of Michigan, Ann Arbor, Michigan, U.S.A.*

Preface

Ecology can be defined as the study of the inter-relationships between organisms and the physical and biological components of their environment. With such a broad field for research, ecologists, perhaps more than any other group of biologists must find it difficult to keep up with developments occurring at the many growing points of their subject. It is the intention of *Advances in Ecological Research* to present comprehensive accounts of selected topics of ecological research in such a way that biologists with a general interest in ecology as well as specialists in ecology, can obtain a balanced picture of what is taking place.

The scope of the articles to be published in this series will not be strictly limited to ecology. Advances within ecology are very dependent on changes occurring in other branches of science. It is important, therefore, that from time to time, articles surveying developments related to ecology in such subjects for example, as genetics, biochemistry, taxonomy and biometrics should be published. There has never been a rigid division between pure and applied research in ecology and many of the major advances in the subject have come from the investigation of applied problems. It is our intention, therefore, that contributions which show the application or development of ecological principles in applied biology should receive adequate treatment in this publication.

This first number contains four contributions which in their different ways discuss important aspects of ecology. Much of our knowledge of the activity of soil organisms is dependent upon more accurate assessments of their numbers and their distribution. The literature of this subject is scattered throughout many journals and the efficiencies of the various techniques are inadequately known. A. Macfadyen's review of methods for the extraction and study of soil arthropods provides information which should be of assistance not only to those concerned with arthropods but to many contemplating or engaged upon the study of other soil organisms. The main theme of M. J. D. Poore's paper is the analysis and description of plant communities. This subject as well as being the basis of plant ecology is of interest to animal ecologists and to applied ecologists concerned with the utilization of biological resources. L. B. Slobodkin in *Energy in Animal Ecology*, deals with a branch of research which is developing rapidly. Here again is a topic which goes beyond the confines of one type of ecologist. The problems of how energy is passed on from one stage in a food-chain to another and whether the efficiency of energy fixation by organisms can be increased are fundamental to the proper use of natural resources. The very long contribution by J. D. Ovington provides an integrated picture of the interrelations which exist between the biological and physico-chemical components of woodlands. This account of the many facets of the woodland ecosystem apart from being a fundamental contribution to ecological science is of special significance to biologists and others concerned with the place of woodlands in land utilisation programmes.

October, 1962 J. B. Cragg

Contents

Soil Arthropod Sampling

A. MACFADYEN

The Method of Successive Approximation in Descriptive Ecology

M. E. D. POORE

Energy in Animal Ecology

L. B. Slobodkin

Quantitative Ecology and the Woodland Ecosystem Concept

J. D. Ovington

Soil Arthropod Sampling

A. MACFADYEN

Department of Zoology, University College, Swansea

I. Introduction

In response to the editor's suggestion this article is intended to provide practical advice on how to obtain efficient estimates of numbers of soil arthropods. The attainable efficiency of any sampling scheme depends greatly on the aims of the investigation, a truism which applies particularly to soil fauna surveys because in such work sampling effort increases steeply with the accuracy demanded. The need for the biologist to balance most carefully his demands in terms of accuracy and taxonomic range against resources of time and equipment is the more acute when he studies unevenly distributed populations, many of them belonging to groups whose taxonomy is, even now, rudimentary and possessing such a wide range of size, density and biological characteristics that the simultaneous sampling of all groups is quite impracticable. Secondly, such a variety of devices for extracting and collecting soil arthropods has already been described (see, for instance Balogh, 1958; Kevan, 1955; Kühnelt, 1961), that the novice is at a loss to know which methods are appropriate to his own work. Some of these are intended, or can be used, only for qualitative work, others apply only to a very limited range of organisms and, despite the excellent reviews mentioned, there appears to be a need for a systematic treatment devoted to the two themes mentioned, namely experimental design and choice of extracting and collecting methods.

It is essential to appreciate, that an experimental design should be conceived as a whole, that the most elegant statistical methods cannot compensate for biased sampling nor for inappropriate application and that the most complete extraction is invalidated by faulty experimental design. Nevertheless it is convenient to discuss the subject under two headings, the experimental design of the survey and the field and laboratory methods to be used, because the first has a wider application and so can be treated mainly by reference to the literature, whilst the latter are more confined in their range of interest to the present topic and therefore demand fuller treatment. In order to relate the methods discussed to practical needs the following section (Section II) suggests a number of types of soil arthropod survey and the appropriate sampling requirements whilst in the final section (Section V) an attempt at synthesis is made by considering some specific projects.

II. Characteristic Sampling Requirements of Some Different Types of Soil Arthropod Survey

Since at least fifty percent of the world's arthropod fauna remains to be described there is still a great need for geographically widespread taxonomic collections to be made without regard to quantitative considerations, especially from regions outside Western Europe. However, our knowledge of the taxonomy of European soil arthropods has improved so much in the last twenty-five years that serious study of ecological problems is now possible because the ecologist can see beyond taxonomic difficulties which formerly occupied his full attention or deterred him from studying the soil fauna entirely. Thus while exploratory surveys can and should be conducted with simple, robust equipment, used on collections obtained in a haphazard manner, ecological work now demands that greater attention be devoted to sampling methods.

Ecological problems are mainly of two kinds; firstly there are those involving comparisons between species lists (sometimes of limited taxonomic range) which have been obtained from different habitats, parts of habitats or at different times, including distributional, seasonal and successional studies. Secondly we have studies that are concerned with trophic structure of communities including food chains, biomass measurements, biological interrelationships and population-size measurements within a single community. The former, which can be called community studies, tend to make greater demands on taxonomic *precision* than the latter or "trophic" type because a greater range of related species is encountered and distinctions between them are often critically important in detecting changes in community composition. In a trophic study confined to a single habitat, fewer species are encountered and general conclusions will hardly be invalidated if an isolated species is misidentified. On the other hand the trophic study may involve a wider *range* of taxonomic work because predators and competitors of different natural phyla cannot be excluded. Further, sampling accuracy is usually more important and sampling difficulties are greater because methods must give accurate information about ecologically related species regardless of taxonomic proximity. Again, the whole purpose of many trophic studies is the establishment of absolute abundance figures and this frequently includes an analysis of age structure as well.

This is not to say that the community-type study lacks sampling problems. Most soil extraction methods operate with varying efficiency in different soils, with the result that quantitative comparisons over a wide range of soil types may even require the use of a series of different techniques. Also, the ease with which even quite closely related groups — for instance chilopods and diplopods or oribatid and parasitid mites can be extracted — varies greatly, so that the use of imperfect extraction methods can seriously upset estimates of their relative abundance. (Macfadyen, 1953, 1955.) However, when allowance is made for such effects the fact remains that the community-type survey does not usually demand that numbers be expressed in terms of absolute abundance on an area or volume basis such as is essential for the trophic study. This is important because counting the fauna of sample-units is often the most time-consuming part of ecological studies and if absolute counts can be dispensed with the ecologist is free to consider alternative measurements such as records of presence *versus* absence and various types of ranking which will be discussed in the next section. For all these purposes a reasonably unbiased extraction method applied to a relatively large number of sample-units can provide more information about community differences in a limited time than a full count of numbers which will necessarily be confined to fewer sample-units. The above remarks are summarized in Table I.

TABLE I

Type of Study	Demands made on Sampling Programme Enumeration			
	Taxonomic	Precision	Covering Different Soil Types	Covering Different Groups
Exploratory	Precise	Unimportant	Comparative	Comparative
Community	Precise	Comparative	Comparative	Depending on range of study
Trophic	May demand broader range Less exacting	Absolute	Unimportant	Depending on range of study

For many groups of soil animals the perfect extraction method remains to be found and this, in the applied field may constitute an insuperable difficulty. The academic ecologist, on the other hand, is usually more interested in principles than in particular instances. He should, therefore, exercise his freedom to choose both the soil type and the animal group with which a particular type of precision can most easily be achieved. Some indication of the relative difficulties associated with the study of different arthropod groups is provided by Table II.

TABLE II

The Amenability of Soil Arthropod Groups for Different Types of Ecological Study

Arthropod Group	Relative Difficulty of Taxonomy	Suggested Method of Extracting from:		
		Litter	Grassland	Arable
Tardigrada	+	A	A	—
Diplopoda	+ +	C	C	CJ
Chilopoda	+ +	F	B	—
Symphyla	+	C	C	H
Pauropoda	+	C	C	J
Woodlice	+ +	C	C	—
Parasitiformes	+	B	B	BJ
Trombidiformes	−	C	C	CJ
Oribatei	+	C	C	CJ
Pseudoscorpionida	+	B	B	—
Aranei	+ +	EC	EC	—
Collembola	+	C	CG	CJ
Thysanoptera	+	BD	BD	DJ
Aphididae	+	CD	CD	DJ
Formicidae	+ +	B	B	—
Coleoptera (larvae)	−	C	C	J
Coleoptera (adult)	+	B	B	J
Diptera (larvae)	−	C	C	J
Insect Pupae	−	O	J	J

Key to Symbols. + + Taxonomy straightforward; + Taxonomy possible; − Taxonomy very difficult or incompletely known; A Wet funnels p. 14; B Dry funnels (fast) p. 16; C Dry funnels (slow) p. 16; D Dry funnels (Chemical) p. 21; E Duffey's extractor p. 20; F Lloyd's extractor p. 19; G Hale's extractor p. 24; H Ladell type Flotation p. 25; J Raw type Flotation p. 25; O None suitable; — Not found in this medium.

The choice of category is, of course, somewhat arbitrary and the fact that a particular extraction method is advocated does not imply that no others can be employed; in particular, as will be shown in Section IV, many potentially useful separation principles remain to be tried and the efficiency of existing methods is not known for many groups.

III. The Design and Execution of a Survey

A. STATISTICAL ASPECTS

The main objective when planning a soil survey should surely be to obtain the required information with a minimum of labour. To achieve this, experiments must be devised in such a way that clear-cut hypotheses are tested and the right number of sample-units is used in relation to the desired degree of accuracy. Both these requirements are usually difficult to meet in practice. In community work one cannot know beforehand which species will occur in a given locality nor how abundant the critical species will be. The number of replicates required to show significant differences is also unpredictable because it depends on total density and on the patchiness of distribution of the animals. In the case of laboratory-based surveys it is possible to carry out a preliminary survey and to obtain rough estimates of density and patchiness on which a rational sampling scheme can be based, but under expedition conditions when the survey is of limited duration there is usually no option but to fix the number of sample-units in the light of previous experience or to take as many sample-units as possible consistent with the labour available. Even here a choice between taking many small or fewer large samples must be made.

When a preliminary survey is possible the mean number m and its standard error s can be calculated in the normal way (see, e.g. Snedecor, 1946). Some idea of the number of samples required in order to achieve a given degree of accuracy can be found (see Snedecor, p. 456 ff.) by substituting in the formula for fiducial limits $m + \frac{ts}{\sqrt{n}}$ where s is the standard error per individual observation and in which t for any given probability level can be obtained from statistical tables.

The question of size of sample-unit, especially when some kind of core borer is used, is often fixed beforehand by the apparatus available. However, there are two statistical considerations involved here, firstly that if the population is distributed in patches, it is essential that the sample-unit area should be small in comparison with patch size, especially if the properties of the patchy distribution are being investigated (see below). If the aim of the survey is to obtain an efficient estimate for a given amount of labour of say the density of a particular species, and if the patchiness of the distribution is only of incidental interest, then the sample-unit area should be such that the ratio of standard error of mean to mean is least. Secondly, it is necessary to balance the advantages and disadvantages of using relatively large numbers of smaller units which usually involves handling less soil but larger numbers of actual cores and pieces of laboratory apparatus. All kinds of practical decisions such as distance from the laboratory and transport facilities are involved here but a useful example of a comparison of this kind is given by Finney (1946) which is also quoted by Healy (in press).

In the example of a wireworm survey conducted by Finney two different sample sizes were tried and in each case the mean and standard error were calculated. By plotting s/m against m and joining the points by eye an expected value of s for any given value of m can be obtained. Since the number of units required to give the same precision are in the ratios of the squares of the s/m ratios the relative advantages of using different sample sizes can be assessed. In this particular instance there proved to be little advantage in using 2in diameter cores rather than 4in diameter ones.

Decisions about sample size will, of course be very much influenced by whether or not abolute numbers are to be determined, that is, whether the survey is of the "trophic" or "community" type as discussed in Section II. Most ecological work done under expedition conditions is of the second kind and is often aimed at detecting and delimiting characteristic species groupings and relating these to environmental factors. Although it has quite commonly been the practice to count all specimens contained in each sample-unit, various expedients can be employed to derive quantitative information from samples without necessitating complete counts. In continental Europe it is usual to estimate numbers to the nearest order of magnitude or to employ an arbitrary abundance scale as has been done by Gisin (1943) and Strenzke (1952) although rather little use is made of these ratings. Presence or absence can be recorded and used to calculate "frequency" (i.e., proportion of sample-units in which a given species occurs). Frequency and mere presence are the basis of correlation tables (or "Trellis diagrams") which are used for community analysis in Scandinavia particularly, for example in the work of Kontkanen (1950, 1957). The recording of the simultaneous occurrence of several species in this way provides a means of determining the extent to which their distributions are associated or complementary. The more modest studies of this kind are content with analysis of correlation between two species at a time, for example, the work of Cole (1949). A recent development however has been the simultaneous correlation analysis of presence or absence of a large number of species by Williams and Lambert (1959, 1960) with the aid of a computer. This involves the complete sampling of a large gridded area and recording species lists for each grid square. The squares are then classified on a hierarchical system which separates those species groups whose distribution is least significantly associated using all possible combinations of species. In this way previously unsuspected correlations between vegetation pattern and environmental factors are detected.

Work of this kind has not yet been done for soil animals but it should be noticed that since only presence is being recorded very small sample-units should be possible — and indeed required, so as to make certain that a proportion of units will be without animals. Under some circumstances this might facilitate taking the large numbers of sample-units which such an analysis demands.

A different approach to community analysis has been suggested — and illustrated — by Fager (1957) who records for each sample-unit the rank order (i.e. the order of dominance) of each species. With the aid of rank correlation analysis developed by Kendall (1955) the affinity of different sample groups can be determined and standard errors can be estimated. Again, the size of sample-units can be considerably reduced compared with those required for complete counts and statistical significance of the results can be established. Also the counting labour can presumably be lessened because a relatively cursory inspection will often suffice to establish order of dominance. When samples regularly contain large numbers of small animals and smaller numbers of large ones, for example oribatid and parasitid mites, it will be desirable to analyse each main group separately.

Sequential analysis [for practical details of calculations see Waters (1955)] is another technique not yet applied to community studies of soil, but which appears to offer substantial savings in labour, and which demands relatively small samples. In cases where a particular species is already established as a valuable indicator and sets of sample-units are being examined for affinity in terms of this species, sequential methods should provide a means of reducing the number of sample-units *examined* to a minimum, although it is unlikely to reduce substantially the number of units taken from the soil in the first place. A somewhat similar method was developed by Capstick (1959) for detecting significant differences between aliquot samples with minimum labour.

A final point to be remembered in connection with the size of sample-units in community studies is that when they are made too small the rarer species are likely to be excluded altogether. It is frequently observed that the commoner species are not those whose distribution is the most useful indicator of faunistic groupings and, as shown by Hairston (1959) it is important to include the less common species. It seems, therefore, that any attempt at a complete community analysis may require either parallel samples with units of different sizes or else the collection of rather numerous sample-units for the rare species and the use of aliquot parts of these or a sequential analysis scheme for obtaining statistics on the commoner ones.

The importance of deciding at the outset whether or not absolute abundance figures are essential should now be apparent. If they are (trophic studies) preliminary samples must be taken to determine the ratio of variance to mean and thus to establish whether distribution is patchy (=aggregated, =contagious, =underdispersed) and, if so, what must be done to measure and allow for patchiness. When patchiness has been demonstrated it is usual to attempt to fit the field data to a theoretical distribution containing terms representing the mean numbers per unit area and the mean size of the patches. The first of these can be used in trophic studies but the asymmetrical distribution invalidates statistical tests which are based on the assumption of a normal distribution and to ignore the patchiness is to discard valid biological information. Distributions which have been fitted to soil sample data include the Poisson, which describes the frequency of random (i.e. non-patchy) events. According to this distribution, which rarely applies in natural soils variance and mean are equal. The negative binomial (see Bliss and Fisher, 1953; Anscombe, 1950) is a distribution related to the Poisson but incorporating the two hypotheses that the population is logarithmically distributed within patches and that these patches occur at random. In addition to m, the mean, the negative binomial uses an extra parameter k which is given by $s^2 = m + km^2$ (when s^2 is the variance). As k approaches infinity the distribution becomes identical with the Poisson whilst as k approaches zero the distribution becomes more clumped.

In practice, as Healy (in press) has shown, k can be determined from preliminary sample data by plotting standard deviation (ordinate) against mean (abscissa) for increasing sample size. The point on the horizontal axis cut by the regression line (drawn by eye) gives an approximate value for k which can then be used with the definitive samples to describe spatial distribution.

Methods for testing the closeness of fit of data to distributions of this kind are discussed by Anscombe (1950), Waters (1955) and Quenouille (1950) while Hartenstein (1961) describes a practical study on aggregated soil arthropod populations.

When data obtained from populations which do not fit a normal distribution are to be subjected to statistical tests such as those used to determine significantly different population levels, the raw data cannot be used because such tests are based on the assumption of normal distribution. In this case the data must be "transformed" by functions which will vary with the type of distribution. Data which fit a Poisson distribution should be converted to their square roots (Snedecor, 1946). When the data fit symmetrically into the groups 0–1, 1–2, 2–4, 4–8, 8–16, etc. logarithms should be taken (Quenouille, 1950) and data which fit the negative binomial should be transformed by $\log(x+k)$.

Taylor (1961) claims that biological data are better fitted to a power series than to the negative binomial. This is given by $s^2 = am^b$ (where s^2 = variance, m = mean, a = constant which varies with sample size, b = constant which varies with, and can be used to measure, aggregation). He suggests tests of significance and claims that b is a biologically valid measure of aggregation.

There remains the physical problem of how the sample-units should be disposed on the ground; whether at random, on a regular lattice, on transect lines or a combination of such patterns. Once again the purpose of the study is the determining factor. A trophic study based on a single experimental area can be gridded and sample-units taken at random. A rather more efficient procedure which still avoids bias is to use stratified random sampling; the whole area being equally divided into sub-areas or "strata" and equal numbers of sample-units being taken at random from each stratum. A linear arrangement of strata can also be used in a community type study when distribution along a catena or transect line is to be investigated, but usually, it is more practicable to select a number of areas along the transect which are gridded and sample-units are then taken randomly from each. Under the crudest field conditions when soil faunas are to be related to floristic or other discernible regions there may be insufficient time to construct grids and to sample at random and in this case the best compromise seems to be to sample at the places where an object thrown into the air lands. In this way unconscious bias is largely avoided.

The completely regular grid has the advantage over random sampling in that efficiency is higher but there is the remote possibility that sample-points will resonate with natural features. Of course when a complete map of spatial distribution is required a rigid grid with sample-units from each square should be used.

To conclude this brief survey of statistical aspects of soil arthropod sampling, it seems appropriate to stress that such methods allow the ecologist who takes the trouble to formulate clear-cut hypotheses to plan an efficient programme of work. There is a wide range of methods appropriate to different types of study and by careful selection the ecologist can not only obtain significant answers with a minimum of labour but also, because he is forced to concentrate on principles he can finally satisfy the natural historian's just complaint that so much quantitative ecology has been mechanically applied and divorced from biological reality.

B. LABORATORY METHODS FOR COUNTING AND RECORDING

Quite apart from questions of sampling in the field and the actual means by which the soil fauna is removed from the sample-units, a good deal of variation exists in the methods by which the animals are counted or examined. A full treatment of this topic is not praticable here, but a few notes may perhaps be of assistance to the newcomer.

Most small arthropod material is ultimately examined in the form of a suspension of animals in specimen tubes (vials) containing a fixative solution such as 80% alcohol. Some arthropods especially Collembola, do not sink readily in such solutions and Gisin (1960) gives a formula for a solution in which they sink more readily. This contains much ether and cannot be used in most Tullgren-type funnels, which necessitates manipulation of the samples to change the fluid. An alternative which the author has found quite satisfactory is to place the corked samples in an oven or other warm place at about 50° C until the fixative has warmed. Usually all Collembola sink under these circumstances but if any are still floating after gently swilling the fluid in the tube they can be sunk by a single drop of ether. Once all arthropods have sunk the bulk of the samples can be safely reduced by decantation.

It is still true to say that no permanent microscopic preparation method exists by which taxonomically critical material can be preserved indefinitely. Probably the best "permanent" methods are the ringed lactic acid preparation of Gisin (1960) for Collembola, the use of Gum Chloral media (see Evans, 1955) for the more robust mites, and of California mountant (Baker and Wharton, 1952) for more delicate ones. But it is far safer to keep whole samples and important specimens in small tubes corked with cotton wool — or better *Polyporus* (Hobart, 1956) and submerged in an outer vessel containing 80% alcohol. Labels are best written in soft pencil on good quality card because some indian inks are not stable. Temporary mounts in lactic acid are then used to examine specimens when required (Evans, 1955). Further details on mounting methods are given by Kevan (1955) and Kühnelt (1961).

A number of quite complicated techniques have been devised for counting submerged animals under the microscope, including the long chamber driven by a gramophone motor under a binocular microscope used by the Freshwater Biological Association for plankton counting and the special nematode counting chambers described by Williams and Winslow (1955) with ruled squares for aliquoting and labyrinths to facilitate searching. Many workers use a ruled, flat-bottomed vessel and search each "lane" in turn. A most useful device is the "gliding stage" (manufactured by W. Watson & Son Ltd) with which any glass vessel can be moved in parallel lines beneath a microscope. Since it is usually necessary to handle specimens and turn over larger species and pieces of detritus, the author finds it more convenient either to push all material to one side as it is counted and as the vessel is moved in parallel lines or, alternatively, to pick out each specimen with fine Swiss watch-maker's forceps. In this way duplication of counts is avoided and no special vessels are required. An office-type tape recorder, operated by a foot pedal, can be an invaluable aid in counting because one's hands are free to manipulate the microscope and forceps and eye-strain due to constantly looking up from the microscope is reduced. When played back the numbers of each species can be counted either on paper or with the aid of a multiple abacus.

All these devices result in economy of counting effort but it should be stressed once more that complete counts should only be employed when the nature of the study renders them essential.

IV. A Survey of Available Examination and Separation Processes

The aim of this section is not to describe complete sampling techniques for all possible situations but to discuss under three main and a number of subsidiary headings the main *principles* which have been employed in attempting to examine or extract soil arthropods. Some examples of the application of each principle, including practical details, then follow and finally a discussion of the inherent advantages and limitations of the employment of this particular principle.

A. The Examination of Soil with Animals *in situ*

Despite its ready availability soil is a remote medium for the biologist to study and many attempts have been made to look inside the soil and see what animals are doing there. Kubiëna (1938) introduced special microscopic apparatus for use in the field, but this has not been widely adopted and the most successful work of this nature has resulted from soil sectioning. Haarløv and Weis-Fogh (1953) developed a method of sectioning soil cores which had been impregnated with agar after the cores had been first frozen in solid carbon dioxide and then fixed in formalin vapour. In this way considerable insight was gained into the feeding behaviour of small arthropods and also into the distribution of animals of different size groups in relation to soil pore size. A different approach to the same problem is that of Alexander and Jackson (1955) who used resin impregnation followed by a grinding technique similar to the methods used by geologists. The Haarløv technique uses simple apparatus and has provided biological information which could not otherwise have readily been obtained: it is however rather laborious and largely limited to commoner species. More recently however Tribe (1960, 1961) has obtained somewhat similar results as a side product of work on microbial decomposition of cellulose. The cellulose was attached to glass slides which were buried and then dug up for examination at intervals. Especially in the later stages oribatid mites as well as nematodes were found feeding on micro-organisms and it seems likely that a related method could successfully contribute to the study of feeding behaviour of small arthropods.

B. BEHAVIOUR-TYPE METHODS

The methods discussed in this and the next section differ from the above in that they are designed to remove the fauna from the soil. This can be done in two fundamentally different ways, firstly, one can take advantage of differences in physical properties such as density and wettability between animal material on the one hand and plant and mineral matter on the other to remove the fauna mechanically. Secondly, one can make use of the animals' own powers of locomotion and response to physical stimuli to attract or repel them away from the soil. Both these principles have inherent weaknesses and neither has yet produced a universal extraction method. Thus the first which are here called "mechanical" methods are more especially restricted to certain groups on account of the great range of density and surface features found in animals and are particularly inefficient at separating animals from large quantities of plant matter. On the other hand they work best in arable soils containing high proportions of minerals, especially clay, under which conditions the second type are least efficient. Behaviour-type methods, as we shall call them also tend to be fairly selective and are useless for removing immobile stages such as eggs and pupae, unless these are deliberately hatched to a mobile condition. They are, on the other hand the only available methods for most media containing much plant matter such as litter and grassland mats.

Considering first the behaviour-type methods it would seem sensible to classify these in terms of the kinds of stimuli such as heat, humidity, and chemical attractants and repellents used to induce the animals to leave the soil. However, in practice most practical methods employ a number of stimuli together and it is more convenient to consider firstly those methods which operate in the field on untouched soil and secondly those which are applied in the laboratory to soil or litter sample-units or cores which have been removed from the habitat.

1. Field Apparatus: Pitfall and Baited Traps

Pitfall traps do not necessarily involve the application of specific stimuli; frequently they depend simply on chance wanderings of surface forms. They can be made very simply from glass jam jars or (less suitably) tins and can be left empty or contain a layer of preservative such as formalin. Many refinements have been described especially from the Kiel school lead by Tischler (see Heydemann, 1958). In damp climates it is desirable to arrange a rain-shedding cover and very dilute phenyl mercuric acetate solution has the advantage over formalin of being non-repellent. (It corrodes aluminium very rapidly however.) An important extension of the pitfall principle is due to Williams (1959) who has devised a clock-driven version which changes the receptacle into which the animals fall at regular intervals; this permits the use of these traps for studies of diurnal activity rhythm.

Baited traps have not been widely used in small soil arthropod work but the interesting results obtained by Walker (1957), who consistently attracted characteristic species to fish, melon and cornflour baits show that they have considerable potentialities. The principle is that of the lobster pot and the structural details are probably not of great significance. The use of meat and fish for catching larger arthropods is discussed by Heydemann.

Obviously the main weakness of all the above methods is that they are not applicable to absolute number estimates and that the catch depends on the activity and behaviour of the animals. Activity is highly dependent on weather conditions and the influence of behaviour is revealed by the finding that many species are never caught in this way whilst there are others hardly ever seen except in such traps, especially those which are baited. However an extended trapping of this kind is particularly useful under expedition conditions for obtaining a good range of species in certain groups, especially beetles, and can be carried on with a minimum of labour and materials.

2. *Laboratory Apparatus for the Treatment of Samples Removed from the Soil*

Although removal of soil to the laboratory has drawbacks such as the alterations caused thereby to the habitat and the possibility of losses and changes in the fauna there appears to be no quantitative method of sampling which avoids these objections. The details of soil core removal naturally vary somewhat with the medium; for instance in hard tropical soils heavy machinery may be needed (Belfield, 1956) whilst sampling of woodland litter may preclude the use of a core sampler of any kind and demand the use of devices such as scissors and carpenter's chisels to cut a mass of material which will fit a tray as nearly as possible (Macfadyen, 1961). Naturally such factors have repercussions on the extraction method used. A large number of core samplers has been described (Coile, 1936; Alexander and Jackson, 1955; Macfadyen, 1953, 1961; O'Connor, 1957). The most usual design consists of a hardened outer cylinder containing one or more removable plastic or metal rings and fitted with a handle. The cutting edge is often made slightly smaller in diameter than the rings and these rest on a ledge some distance above the cutting edge. This results in the loss of the deepest part of the soil core, an unimportant matter in deep soils but a serious difficulty in shallow ones. In the latter case the author has obtained satisfactory cores by sinking a plastic ring into the ground with the aid of a carpenter's chisel. A disadvantage of any system which uses rings of fixed depths is that natural boundaries of the soil profile are ignored and any vertical separation which is done by cutting the soil core at the ring junctions is quite arbitrary. This difficulty is overcome by O'Connor (1957) in his longitudinally split sampling tool. It has been argued that serious losses can result from compression caused by the use of core samplers (Murphy, 1958) but there does not appear to be any quantitative evidence for such an effect in the case of the type of sampler just described. It does appear, on the other hand, that manipulation of the soil core by hand can lead to reduced yields (Macfadyen, 1953) and doubtless losses would occur if the soil were seriously compressed. Most workers now agree that an intact soil core yields more animals than a teased-out one and that a core from near the surface should be inverted when placed in the funnel-type extractor so that surface living animals with limited digging powers can easily make their way out (Macfadyen, 1953).

i. Wet Funnel Methods. These are suitable for extracting animals which normally live in the water films of the soil and have been widely — and successfully used for such groups as nematodes, enchytraeid worms and rotifers (e.g. Nielsen, 1948; Peachey, 1959; O'Connor, 1955). Among the arthropods only the Tardigrada are water film dwellers and they can certainly be obtained in large numbers from moss and similar media by means of the Baermann funnel. This is simply a glass funnel fitted with a pinch cock below and a wire gauze some distance below the rim. A sample is placed on the centre of the gauze leaving a clear annulus round the edge and the whole is flooded with water and placed beneath an electric bulb of such a wattage that the temperature of the water near the sample rises to about 40° C in two hours. As the temperature rises aquatic animals swim actively into the water and then succumb to the higher temperature and sink to the stem of the funnel. After about twelve hours the pinch cock is opened for long enough to draw off about a quarter of the contents of the funnel. The process can be repeated.

An inverted wet "funnel" method in which the animals move upward from warm to cold temperatures and into a layer of sand, from which they are removed by simple flotation has been used for enchytraeids (Nielsen, 1953) but not for soil arthropods; conceivably it could be of value for collecting small semi-aquatic crustacea such as harpactids, which occur in certain soils.

ii. Dry Funnel Methods. For fuller accounts of the history of these methods see Heydemann (1958) and Macfadyen (1953).

Berlese (1905) was the first biologist to use heat in conjunction with a funnel for collecting small arthropods. He used a water jacket round the funnel and the sample was suspended near the top so that it was dried out and animals were forced to leave the medium largely through desiccation. Berlese was not concerned with quantitative results and made no attempt to test the efficiency of the apparatus nor to apply any kind of directional stimulus. Tullgren (1918) introduced the use of an electric light bulb suspended above the funnel in place of the water jacket largely as a matter of convenience and economy; there appears to have been no intention to produce a heat gradient by this means. A large number of later authors have introduced variants on Tullgren's funnel often with particular purposes in mind. Some of these are as follows.

Ulrich (1933) Murphy (1955) Kempson, Lloyd and Gelhardi (p. 19)	Devices to reduce fall of soil into sample container.
Forsslund (1948)	First comparative trials on effects of sample size on yield.
Ford (1937)	First author to group many small Tullgren funnels in a single apparatus for purpose of applying statistical methods. Used black heater in place of light bulbs.
Haarløv (1947)	Showed that condensation in funnels reduces yield and introduced measures to prevent this.
Macfadyen (1953, 1955, 1961)	Introduced principles of taking samples in cylindrical tubes, of establishing a deliberate steep temperature gradient (1953) and later humidity gradient (1955, 1961), thus combining repellent and attractant stimuli.
Duffey (Private communication)	Special high-speed horizontal extractor for spiders. (See p. 20.)
Murphy (1955) Auerbach and Crossley (1960)	Other multiple Tullgren funnels.
Kempson, Lloyd and Gelhardi	Use of intermittent infra-red heat. (See p. 19.)

The precise type of apparatus used depends on the completeness of extraction required, the medium to be investigated, the sources of heat available and the groups of animals being investigated. Naturally, large funnels are required for the larger, less numerous animals and if only general collection is the aim, very simple apparatus such as collapsible funnels made of paper or polythene, suspended in the sunlight will yield numerous animals. For more quantitative yields of larger species, the rate of heating and humidity status in the funnel beneath the sample become important. In the case of compact humid soils of temperate regions sufficient moisture can be derived from the soil itself and humidity can be controlled by regulating the draught through the funnel (Macfadyen, 1961).

In work on micro-arthropods (mites and Collembola) sample-units of about 25 cm^2 area will usually yield high enough numbers for complete counts and 10 cm^2 for community studies. If counts are not to be biased in favour of certain groups (as certainly happens in the normal Tullgren funnel operated in dry conditions) it is necessary to arrange that humidity beneath the soil is high and the temperature low. These aims can be achieved in the canister type extractor (Macfadyen, 1961), by standing the canisters, into which the animals fall, in a water bath through which cold water is circulated. Less accurate work, on the other hand, with less elaboration, can be done with simpler extractors of the kind described by Macfadyen (1953) and Auerbach and Crossley (1960). Frequently the medium to be sampled will not permit the digging of cylindrical plugs of soil such as are required for the above, because of the presence of sticks in litter and of stones in soil. In this case the canister extractor cannot be used because the samples allow too much drying from above and positive steps must be taken to ensure humid conditions beneath the sample. This calls for an air conditioning system such as is incorporated in the 100 cm^2 funnel extractor of Macfadyen (1961), an expedient which can only be realized in a properly equipped laboratory. This extractor, which can, of course, also be used for soil sampling requires careful insulation if it is to be operated in a heated room and a supply of cooling water (or a refrigerator) which can extract about 2 cal of heat per minute. The simplest form of construction seems to be a slotted angle steel frame with hardboard outer shell and insulating panels of expanded polystyrene ("Jablite") placed between the frame. Full details are given in the above reference.

There is a growing tendency towards carefully planned community studies in rather inaccessible places and consequently there is an urgent need for an apparatus with which fairly complete extractions can be achieved without the aid of electric current. In an effort to meet this need two extractors have been made recently, the first using paraffin (kerosene) as a source of heat and operated for a summer season in Lapland, the second using bottled-gas (butane — or pentane in cold conditions) which has not been tried in the field at the time of writing. Both these extractors are similar to the canister extractor (Macfadyen, 1961) in general layout and, unless ambient temperatures are very low, both require a flow of cold water for cooling purposes (usually fairly easily arranged from a stream under expedition conditions). The paraffin-operated extractor contains a hot water-bath (in place of the electric heater) which is supplied with a circulation of hot water from a separate boiler heated by a wick type boiling stove ("Valor 65 S"). The flow and return pipes should be of thick transparent PVC tubing (e.g. E.R.P. tubing from Esco Rubber Ltd, London), and these and the boiler insulated with glass wool. An apparatus to deal with forty-eight sample-units consumed about 3 gal (14 l) of paraffin per week. The main problems under field conditions arose from the difficulty of ensuring that the paraffin heater worked efficiently (with a blue flame) and without boiling the water. In other words a smaller stove run with a higher flame would have been easier to control and might have burned less paraffin.

The gas-burning extractor (Fig. 1) uses a $\frac{1}{8}$in thick aluminium hot plate in place of the rather cumbersome boiler and water circulation and is, therefore, more compact. The gas burners are incorporated in the sides of the apparatus which are lined with heat-resisting plastic (Tuffnol, Asp Brand). An apparatus to treat thirty samples burns about $17\frac{1}{2}$ lb (8 kg) of butane per week. The relative advantages of burning butane, at about two and a half times the price of paraffin but weighing about the same in cylinders are largely a matter of expedition logistics against the greater convenience of the gas.

If the gas is to be kept at or below 0° C propane should be used instead of butane; this has about 75% the heat of combustion of butane and is more expensive. The entire apparatus including core sampling tools, canisters and rings weighs 40 lb (18 kg).

The use of vertical-sided canisters in place of funnels in extracting small soil cores has the advantage of removing all possibility of loss of animals on the sides of the funnel, of reducing the chances of loss due to worms, spider's webs and other obstructions in the constriction of of the funnel and of presenting a large aqueous surface for the maintenance of a high humidity beneath the sample. Two corresponding disadvantages are that removal of material from the canisters requires more skill and larger quantities of fluid than in the case of a vial on the end of the funnel and also that the large liquid area necessitates the use of a non-volatile fluid to ensure that animals will not be deterred from leaving the sample. With very little practice the "catch" in a

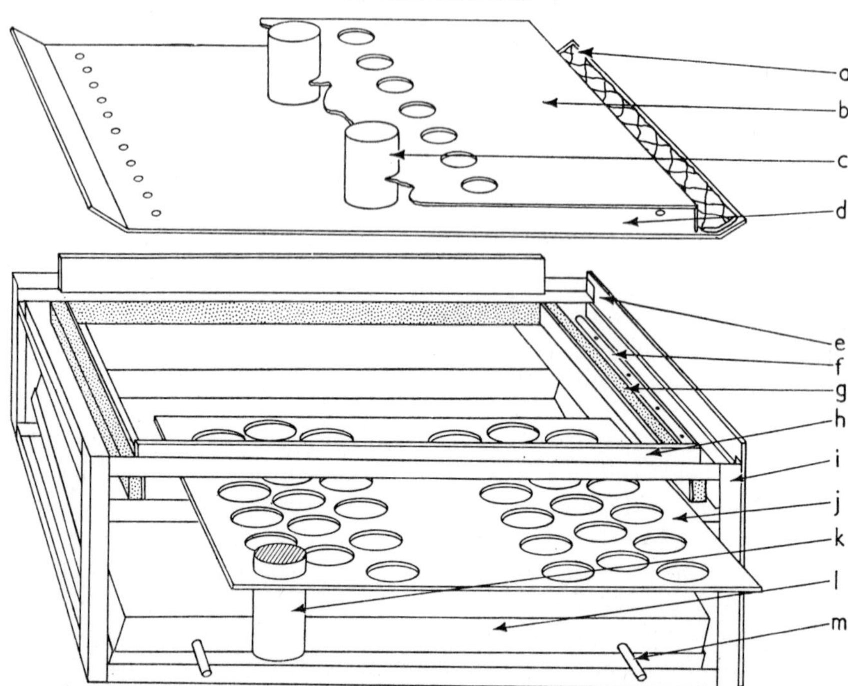

FIG. 1. Canister-type extractor heated by bottle gas for use in the field. a. Strip of asbestos tape for insulation. b. Tuffnol (laminated phenolic plastic sheet) cover to conduct pre-heated air to samples. Cold air enters by central holes near chimneys, which are also used for heating vials to make Collembola sink. It passes to the hot chamber via the small holes at the edges. c. Chimney to evacuate damp air. d. $\frac{1}{8}$ inch-thick aluminium hot plate. e. Tuffnol draught shield. (Similar shield on left side removed for clarity.) These should be "Asp" brand. f. Gas heater consisting of copper pipe with five Primus nipples. g. Foamed polystyrene ("Jablite") insulation to hot chamber. h. Tuffnol strip to complete cover. i. Slotted angle frame. j. Tuffnol rack supporting soil samples and canisters (by means of rubber bands). k. Aluminium canister. l. Aluminium cold water bath. (In lowered position — the frame is hinged at the back.) m. Pipe for water circulation in cold bath.

25 cm² area canister can be swilled and washed with alcohol into a 3 in × 1 in (75 mm × 25 mm) specimen tube, but it is desirable that the samples should be further reduced in bulk to a volume of a few cm³. Further, the use of the aqueous media needed to comply with the second requirement means that most Collembola are still floating on the surface at the completion of extraction. The author's practice is to place the 3 in × 1 in vials containing the catch plus water plus alcohol in a warm place until the temperature has reached about 45° C. Usually all Collembola sink when the medium is gently swilled round under these conditions, but if not, a drop of ether or a special degreasing fluid as advocated by Gisin (1960) can be added. After a further period of settlement nearly all the fluid can be safely decanted and the contents

of the vial then washed into a smaller tube for storage. Admittedly this procedure takes more time than the use of alcohol in a tube at the bottom of the funnel, but in compensation one is assured of (i) a non-repellent catching fluid (a large surface of alcohol is distinctly repellent to some groups; see Macfadyen, 1961), (ii) a fixative fluid of controlled strength and composition in the storage tubes (according to Gisin the degreasing fluid should be replaced by a different fixative solution in any case) and (iii) the high humidity conditions beneath the sample which are necessary for efficient extraction.

There remains the question concerning what aqueous medium should be used in the canisters. The latter must be odourless, non-repellent and non-volatile, should preferably kill animals falling into it to prevent predation and escape, should kill moulds which readily attack dead arthropods and should not corrode the containers. For one or other of these reasons all the previously used fluids are excluded. A rather unsatisfactory compromise was found in "Agrimycin", a fungicide which was advocated in the above paper. Since then a slight repellent effect has been demonstrated and, more seriously, it has been shown that earthworms in this solution produce a copious mucus secretion and undergo autolysis which can make the arthropods very difficult to separate. Recently Lloyd (see below) has described the use of a picric acid solution and tests by the present author show that one part saturated picric acid to one part water has no detectable repellent effect and completely fixes all animal tissues. The yellow staining may sometimes be a disadvantage in the determining of new material, but at present this appears to be the best medium available. Tests on further possible substances are continuing.

A new device for extracting arthropods from leaf-litter has recently been developed at the Bureau of Animal Population, Department of Zoological Field Studies, Oxford University, by Denys Kempson, Monte Lloyd and Raymond Ghelardi. (A detailed report for publication is in preparation, but the following brief note has been authorized for inclusion in the present article.) Each sample is suspended a short distance above an aqueous solution of picric acid contained in a straight-sided closed vessel. The collecting vessels are immersed in a circulating cool water-bath that is well insulated from the infra-red heat source above the samples. A battery of infra-red bulbs switches on and off in alternation with conventional light bulbs, producing an adjustable pattern of heat pulses with continuous light. A high humidity is present directly under the sample from the beginning. The extraction process requires about a week, during which time very high gradients of both humidity and temperature are established through the samples. Extraction efficiencies have been determined by soaking and hand-sorting the dried residue and by introducing known numbers of organisms. Litter copepods, nematodes and enchytraeid worms are extracted with very low efficiency — of the order of 10% or less. Mites, Collembola and most other arthropods are extracted very efficiently — of the order of 90% or more. Snails and fly larvae, as a group, are extracted with considerably less than 90% efficiency. Each sample is spread over two layers of cotton fillet net, whose openings are filled with delicate linty fibres. These tiny fibres offer no serious impediment to the passage of animals but do hold back most of the litter debris, resulting in a much cleaner extraction than is usual with Tullgren work — and a great saving in counting time.

This apparatus resembles the "canister" extractor in using straight-sided water-cooled containers and an aqueous medium for the purpose of establishing high humidity; it is also the origin of the use of picric acid fixative. The main differences are in the use of the cotton fillet net which should show a distinct improvement in the cleanness of samples, and in the use of a pulsating heat source in conjunction with light. The advantages of this form of repellent stimulus will, no doubt, be justified in the forthcoming paper.

A special extractor has been developed for obtaining spiders from soil and litter, mainly in the form of undisturbed turves, by E. Duffey, of the Nature Conservancy, to whom the author is greatly indebted for an account of the apparatus. This exploits the greater mobility and better-developed senses of spiders, which are driven horizontally out of the medium by a powerful electric-fire bar, the whole extraction occupying only a day or so. The animals are collected in a trough containing dilute glycerine and pairs of spider-tight chambers are conveniently arranged on either side of fire bars with the troughs on the outside. There is some difficulty, due to condensation, in handling wet samples and the efficiency of the apparatus has not yet been exhaustively tested.

In most of the Tullgren-type extractors mentioned so far, heat, light and desiccation in various combinations have been used as repellent stimuli and in some, low temperatures and high humidities have been added as attractants. There remains a very real need for more fundamental behavioural work on the various groups of arthropods with the ultimate hope that at least some sorting out of fauna will be achieved by submitting samples to gradients of different factors: this is a long-term matter and beset by many technical difficulties (see Madge, 1961; Macfadyen, 1962) and clear-cut behavioural responses are often hard to demonstrate. At present the most hopeful approach seems to be the simultaneous application of as many repellent-attractant gradients as possible, and certainly the use of combined heat and humidity gradients has proved rewarding. The role of light in the Tullgren-type extractor has never been satisfactorily determined but, since light hardly penetrates any appreciable depth into soil, it seems unlikely that it will do more than deter animals from escaping upwards at the start of an extraction. Once the heat gradient is established upward escapes seem unlikely and are not encountered in practice.

Since most invertebrates are sensitive to chemical vapours (see Dethier, 1947) it is rather surprising that these have not been more widely used for selective extraction of soil fauna. Macfadyen (1953) showed that higher yields of soil aphids and thrips could be obtained by simply enclosing one end of a soil core in an atmosphere containing dimethyl phthalate and suspending the other end over an open funnel so that a vapour-concentration gradient would result. Since then more effective repellent substances have been marketed for use against biting flies and it has also been shown that a dry atmosphere is far from an ideal attractant. It seems likely that better results on a wider range of organisms could be obtained by exploiting these two advances and possibly a very simple method could be worked out for field use.

In conclusion, Tullgren-type methods have the great advantages that they demand relatively simple apparatus, can be used under field conditions, operate well even in the presence of much plant matter and require little labour on the part of the operator. Near-perfect yields can be achieved in particular cases at the cost of more complicated apparatus and greater labour. On the other hand the variability of behavioural responses shown by soil arthropods renders it unlikely that a single Tullgren-type extractor can yield complete extractions of all groups and the fact that such methods rely on the animals' own mobility implies that only mobile stages can be obtained in this way.

C. MECHANICAL SEPARATION METHODS TAKING ADVANTAGE OF DIFFERENTIAL PHYSICAL PROPERTIES OF ANIMALS AND THE SUBSTRATUM

This group of extraction methods is based on mechanical means of separation, exploiting differences of physical properties between the bodies of animals and of their surroundings. The powers of movement of the animals are ignored so that immobile stages such as eggs, egg capsules and pupae may be extracted — and indeed there is no other way of extracting them apart from waiting for them to hatch into mobile stages. A not very important, compensating disadvantage is that dead organisms will also be extracted and may be confused with live animals. As with behaviour-type methods it is hardly to be expected that a single mechanical method will separate all the fauna from the soil. The reason for this is that animals vary in surface properties, density and size and in all these respects are likely to overlap inanimate materials, especially in the case of those mites, opilionids and insect pupae which habitually coat themselves with mineral and vegetable particles. In this case it seems most logical to classify the types of methods according to the physical principles used despite the fact that two or more of these are frequently combined in a single extraction programme.

1. Sieving

Sieving is hardly likely, by itself to provide a means of separating a wide range of animals from soil, but it is often a most useful adjunct to other methods either for removing inanimate material outside the size range which is of interest or for separating animals — perhaps a single species — which is of economic interest. An example of the first technique is the use of a dry power-driven sieve such as that described by Lange *et al.* (1955) and also used by Lloyd for removing leaves in the litter fauna study mentioned above; in this case all the smaller fraction is then treated by the special behaviour-type method — which could not have been applied to material containing large leaves. It is more usual, however for sieving to be done under water, often with the aid of powerful jets. This was the basis of Morris's (1922) wireworm technique in which a battery of sieves was used, stones and roots being retained by the coarser meshes and fine silt and clay passing through the finer ones. It was shown by Ladell (1936) that this method by itself, is both laborious and inefficient and he supplemented the use of sieves with flotation on magnesium sulphate (see below). Both Ladell's and later flotation methods have incorporated sieves for the purpose of primary separation of larger and smaller material.

It is also profitable to sieve the material in alcohol obtained from Tullgren-type extractors under certain circumstances, especially when it contains appreciable quantities of large plant material or large numbers of animals. In this case by using mesh sizes of about 1 mm and 200 μ the counting can be done in two or three stages and removal of debris of very different sizes leads to greater accuracy.

2. Sedimentation

As an alternative to separation on the basis of size there is a number of possible methods of separation by density. In most cases these can work on either static or dynamic principles, for example fluids of different densities can be used so that bodies of corresponding densities also come to rest at definite boundaries, or differences in rates of sinking can be exploited to achieve separation. Remarkably little effort has been made to apply these principles to the separation of arthropods although a number of preliminary investigations, not without promise, are known to the author. Dr A. J. Nicholson has described, in a personal communication, the successful use of a tall glass tube containing sugar solutions of increasing density for the separation of blow-fly resting stages from frass. The column is filled in stages, using a disc to prevent mixing, and the un-separated mixture poured in from the top. Different materials separate at different heights and by tapping the column carefully from the bottom can be drawn off in turn. An example of a dynamic system is Ostenbrinck's (1954) (see also Williams and Winslow, 1955) use of a tall vertical plastic funnel arranged with its wide opening upwards for the separation of nematode egg capsules from soil. A constant flow of water is arranged so that it enters the funnel from below and overflows over a lip at the top. Soil, containing capsules is emulsified with water and introduced in an equal stream about halfway up the funnel. The rate of flow is highest at the bottom and least at the top so that particles are suspended in horizontal layers by the upward flow according to their rate of sinking; mineral particles are removed at the bottom. In Ostenbrinck's technique, since only a single kind of object is to be separated, the flow rate is arranged to carry the capsules over into the sieve continuously. However it would appear possible to develop this device as a means of separating objects of different sinking rates. Floating materials could be passed over into a sieve from the lip at the top by progressively raising the rate of liquid flow. The reverse process of separating materials according to rate of sinking in a long vertical tube and collecting them in a moving trough might well offer a means of separating small animals from soil and from each other.

Kühnelt's use of the centrifuge (1948, 1961) is almost the only sedimentation method that has been operated successfully as a complete system with arthropods. The crude sample is centrifuged in concentrated calcium chloride solution and the animals decanted from the surface. It is claimed that by repeated centrifuging with reduced concentrations of calcium chloride solution much of the plant debris, as well as mineral matter, can be removed. However it is clear from the description that great care is required in operating this technique and that separation is not perfect.

3. *Differential Wetting*

Salt and Hollick's (1944) extraction method (see below) appears to have been the first to make use of the differential wetting properties of insect cuticle and of plant matter. By shaking a mixture of the two in a vessel containing paraffin (kerosene) and water it is possible to cause the insects, whose cuticle is wetted by the paraffin, to separate from the plant matter (which is not) when the two liquids are subsequently left to re-form as two immiscible layers. Subsequently Raw (1955) introduced the use of benzene which has a freezing point above that of water. By standing the vessel containing the two liquids in a refrigerator and freezing the benzene (melting-point 5° C) the latter could be picked off as a solid lump and then melted; the animals being easily freed in this way from particles in the water fraction. The paraffin-water interface technique has also been used by Davies (1955) in conjunction with a motor-driven "vacuum cleaner" for separating arthropods from litter in the field. A stream of air carried the animals and mineral matter from the soil and projected the material at the two liquids; this caused everything to be wetted, the arthopod-containing paraffin layer was retained and the plant litter and soil which came to rest in the water layer were rejected.

Undoubtedly the principle of differential wetting at an interface has enormously increased the efficacy of mechanical extraction techniques. It is not, at present, a universal method because many kinds of arthropods, among them woodlice, myriapods and some groups of mites, are not wetted by the chemicals used. Furthermore even the groups which are wetted are inefficiently extracted from large proportions of plant matter presumably because the animals cling tightly or are hopelessly entangled (Evans, 1953).

The use of these methods has, therefore, been mainly confined to separation from arable soils containing much mineral matter and especially clay. Under these circumstances, however, the funnel methods are least efficient and the two types of technique thus complement each other.

In the field of mineral separation the art of controlling wetting properties of different metals has been developed to a very high degree and by the use of surface active substances under controlled conditions of salt concentration and pH, given minerals can be made to float or sink at will. It seems that this is a field in which further experimentation based on a methodical plan could greatly improve the effectiveness of separation processes for biological material and perhaps also permit the sorting of different groups of animals.

Some groups of arthropods are particularly prone to float even in pure water as a result of wax layers and outstanding among these are the Collembola. W. G. Hale (to whom the author is greatly indebted for an account of his method, which is to be published shortly) has developed a method for extracting Collembola from peaty soils employing the principle that vegetable material sinks in both water and magnesium sulphate solution, when boiled at room temperature under reduced pressure. About half the Collembola remain floating on the surface, due to the hydrophobic nature of their cuticles, and the rest may be brought to the surface by subsequent stirring and bubbling air through the liquid.

Evidently much further progress is possible in the field of separation of arthropods by exploitation of differential wetting properties.

4. Flotation

Flotation is a complex process which incorporates both sedimentation and differential wetting to extents which have not always been clearly appreciated. It was first introduced by Berlese (1921) in a simple apparatus described and tested by Balogh (1938), which consisted of a boiling tube with a constricted neck and a plunger. Material from the "Berlese funnel" was mixed with salt solution by agitating in the tube. When animals had floated to the surface the plunger was raised and the surface solution poured off, leaving mineral matter behind. This principle was developed further (independently?) by Ladell (1936) for wireworm extraction from arable soils. After washing through a series of sieves, in the same way as Morris had done, the material of intermediate size was mixed with magnesium sulphate solution and subjected to stirring and bubbling. Plant and animal matter were carried upward, floated over a sedimentation tank and retained by a sieve. The precise role of the bubbling seems rather uncertain and was, perhaps, more concerned with stirring the material than achieving separation by differential wetting. The same principles were then used by Salt and Hollick (1944) for extracting smaller arthropods and by Raw (1955) who added in turn the important stages of separation at an interface (mentioned above) and of pre-treatment with sodium hexametaphosphate ("Calgon") followed by subjection to a vacuum pump and then freezing. This has the effect of breaking up clay soils and liberating the animals much more readily from the mineral material when the samples are subsequently washed.

As flotation processes these methods all seem to depend primarily on the differential density effect of mixing with magnesium sulphate solution and a combination of floating with differential wetting such as is used in mineral dressing does not seem to have been attempted.

Simpler flotation techniques are often useful for limited applications. Thus Edwards (1958) has described the use of plain water and of magnesium sulphate solution in a beaker or other container to obtain Symphyla in large numbers from arable soil. Similarly Nielsen (1953) in a special extraction method deliberately drives enchytraeids into sand of uniform particle size and then separates them from it by simple washing. A similar technique has been used for extracting mud-living harpactids.

Flotation methods obviously have a number of distinct advantages for arthopod extraction. They do not depend on the animals' behavioural characteristics and are particularly suitable for the treatment of deeply worked arable soils containing little plant matter. They also have the advantage of flexibility of sample-unit size: the same extraction apparatus can be used for treating sample-units of varying sizes and there is no objection to increasing or decreasing the size according to statistical requirements as indicated in Section III.

V. The Choice of Complete Sampling Systems for Specific Purposes

It has been repeatedly stressed that a sampling system needs to be planned as a whole. Usually one or two main features of an investigation are irrevocably determined by the purpose of the whole study, for example that the population size of a particular species is to be quantitatively determined in a limited locality or that the species spectra of a large number of sites are to be compared and related to botanical and topographical features of a large area. In particular, the number of species and of higher taxonomic categories, the exactness of quantitative estimations, the nature of the soil and the range of soil types can all be studied either intensively or over a broad range. Wherever any of these requirements give any latitude for choice the experimenter should, naturally, relax the technical demands as much as possible. If the groups to be studied are unimportant the easier ones for taxonomy and extraction (as indicated in Table II) should be chosen. If the exact location is unimportant then places with relatively few species and with uniform light soils will naturally be chosen. If the study is of the nature of a community survey, time will not be wasted over exact counts from large samples.

As regards the choice of extraction methods, the advantage of the different principles have already been discussed under the appropriate headings, and the applications of particular treatment methods are briefly summarized in Table II.

In order to illustrate more specifically what the above precepts imply three complete sampling schemes will be briefly discussed.

1. An investigation into the effects of different artificial treatments on the small arthropods of rough grassland.

The statistical aspects of this kind of problem, which is well suited to demonstrating clearly the effects of simpler environmental factors are well known. The treatments are applied to a latin square or randomized block design and their effects are determined by the analysis of variance technique (Snedecor, 1946) or by a partial regression technique (Satchell, 1955). The blocks subject to different treatments are gridded and regular sample-units taken from each block on a random basis. The size of sample-unit to be used will depend on the density of the animal groups under study. In this case, parasitid mites for instance, which are often quite distinct indicators of the environmental condition, might be considered and, since they might be expected to occur at average densities around one or two per cm^2 for all species a sample-unit of surface area not less than 25 cm^2 is indicated. On the other hand the deliberate taxonomic limitation of the study to oribatid mites with densities approaching 20 or more to the square centimetre would permit sample-unit size to be reduced to, say 10 cm^2 or even less. These sizes are chosen assuming that complete counts are demanded, but if the purpose of the study is to detect general effects of treatments on the relative importance of the different groups, the analysis could be done on the results of ranking the different species in the sample-units. This technique could well be satisfied with sample-units only about a quarter of the size but might demand an increased number of units to obtain statistically valid results. The methods for determining the number and size of units to be used in the case of a complete count have already been given. The use of an analysis based on ranking would demand perhaps sixty-four sample-units or more (4 treatments × 4 replicate blocks × 4 units per block) on each sampling occasion. The actual weight of soil would be small but the number of units obviously makes great demands on the labour of handling. This, and the superiority of behaviour-type methods in grassland studies (owing to the presence of much plant detritus) all point to the use of a Tullgren-type extractor, and for groups other than adult oribatid mites the canister type is indicated if some groups are not to be biased at the expense of others.

There remains the question of depth of sampling. Preliminary investigations would be required here, but in most rough grassland soils the fauna is markedly concentrated near the surface. Unless large numbers of deeper forms such as rhodacarids, protura and *Tullbergia* species were found, sample-units of 3 cm depth would probably be adequate. Otherwise either a proportion of sample points or even all of them might require a second unit from 3 cm to 6 cm depth. In the latter case, of course the number of sample-units would be doubled again.

2. An investigation into the changes in the total arthropod populations of a single arable field under different crop régimes and over an extended period.

This is a further example in which absolute abundance figures are not required. The main difficulties here are the varying nature of the soil under changing cropping régimes and the demand for a knowledge of the complete arthropod fauna. On the assumption that the management would be largely arable and that grass crops would be in the form of leys, not producing a dense turf, Raw's modification of the Salt and Hollick extractor is indicated for the small arthropods. The field as a whole would be stratified and periodical samples taken on a random basis. It is perhaps worth stressing here that considerable seasonal variations in arthropod numbers do occur (Boness, 1953; Dhillon and Gibson, 1962; Macfadyen, 1952) and that quite frequent sampling occasions (perhaps bi-monthly) would be necessary to avoid false conclusions. Once again the use of dominance order or of ranking coefficients would probably save much labour in a study of this kind, although it must be admitted that such methods do not yet seem to have been applied to this type of work. A preliminary investigation would certainly be required to determine the optimal size and number of the sample-units. The flotation method has the advantage that quite a variety of sample sizes can be treated and it might well be worth while to take two or more sets of sample-units of different sizes and to treat them in flotation chambers with correspondingly different mesh sizes. In this way the larger but less common species, up to at least the smaller species of beetles, would be extracted in adequate numbers. There would remain, however, the quite sparse species, mostly predators, such as the large ground beetles and the spiders. In a non-quantitative study of this kind it might be best to sample these with the aid of pitfall traps *provided* these were operated almost continuously over quite long periods. This is necessary because the catches in such traps vary very much with weather and other conditions and also because most of these animals have marked periods of seasonal abundance (see for instance Van der Drift, 1959). Obviously the different groups sampled by these separate methods would be treated in parallel and the changes of species in each major group would be mainly of significance within that group.

3. A trophic study of a single species or a limited group of species in a particular soil fauna. No particular species specified.

This is a class of study which has been very little attempted so far; work includes an investigation by Clark (1954) into the Amphipod *Talitrus sylvaticus* in an Australian rain forest, and the Durham studies on the Pennine peat fauna (see Cragg, 1961). The object of the work is to gain insight into the trophic importance of particular species by measuring the seasonal changes in abundance, age structure and metabolism. The demand for a truly quantitative census limits the range of possible species but our ignorance is at present so profound that the investigator can afford to choose in this respect and his attention is likely to be attracted by species which are shown, in a preliminary survey, to be abundant. The choice of size and number of samples will be made in accordance with the principles discussed in Section III.

The most acute technical difficulties arise when such studies are to be based on woodland fauna. Most woodland arthropods are present both in the litter (Förna) and in the F and H layers beneath it; some species such as the Onydhiuridae (Collembola) are confined to the soil and others such as spiders and perhaps the larger Collembola inhabit the litter, provided it is present throughout the year. It follows that to confine the study to true soil species would limit greatly the choice of species and at least preliminary investigations should cover both litter and soil. The high proportion of undecomposed organic matter in the F layer would usually point to the use of a funnel-type apparatus of advanced design for the soil fauna investigation but the normal Tullgren funnel is unsatisfactory for most types of litter and careful tests of the relative advantages of the Lloyd, Kempson and Ghelardi method against the small funnel extractor with air conditioning would be desirable before coming to a positive decision on the best means of extracting litter fauna. The vertical division of the medium into litter and soil layers is obviously important if the distribution of the animals is to be followed but it is imperative to use the same sampling points for obtaining both the litter and the soil units if errors are to be avoided. For this reason it might be preferable to use the same technique for both soil and litter samples even if a vertical separation proved necessary. It remains true that methods for the satisfactory separation of arthropods from woodland litter have still to be tested and developed.

VI. Summary and Conclusions

This paper is intended as a practical summary of current methods used for sampling soil arthropod populations. Broadly speaking such methods are used in three main types of work: exploratory work which is not concerned with quantitative population estimates, but is aimed at finding what species occur in an area; secondly "community" studies which are concerned with the relative abundance of a wide range of species and often over a wide range of habitats; thirdly "trophic" studies which demand a knowledge of absolute abundance usually of relatively few species and in a single habitat. Each of these fields of work makes its particular demands on sampling and extraction techniques particularly regarding the sizes and numbers of sample-units, the accuracy expected and the range of organisms to be found and enumerated. This has not always been recognized in the past and labour and resources have been inefficiently employed both in the field and the laboratory. The planning of a complete sampling system requires careful choice and matching of experimental design, extraction technique and methods of examination and counting in the light of the above considerations.

The third section is concerned with the special problems posed by the patchiness of distribution which is usual among soil animals and the application of ranking and other alternatives to a complete census in community studies which do not necessitate the great labour of complete counts.

The fourth section is a systematic survey of the main types of collection and extraction processes employed in the study of soil arthropods. Three main types are recognized, those which are intended for the study of animals within the soil mainly by means of sections, the "behaviour" methods in which animals are induced to leave the soil by attractant or repellent stimuli and the "mechanical" methods by which animals are removed as a result of differences in physical properties from the surrounding medium.

Among behaviour-type methods the use of pitfall traps and baited "lobster pot" traps can be used in the field whilst various devices developed from the Berlese funnel and the wet or Baermann funnel demand that the samples be returned to a laboratory for extraction.

Mechanical methods are based on sieving, sedimentation or differential wetting and flotation. These principles can be used successively and material from behaviour methods can also be cleaned mechanically. In general, the behaviour-type methods have the disadvantages, when compared with mechanical methods, that resting stages (eggs, pupae, etc.) are not removed, that, in any one apparatus the size of sample-units is often fixed, and that changes in the populations of animals within the samples can occur during extraction due to hatching of eggs or to predation.

On the other hand the behaviour-type methods have certain clear advantages. Sorting labour is reduced (because the animals walk out of the samples by themselves) a factor of especial importance when many small samples are to be treated. The simple behaviour-type apparatus is cheaper and better adapted to field use, the avoidance of sieving presents loss of the very numerous small forms such as Prostogmata, Scutacaridae etc., and there is much less damage to specimens due to the more gentle handling of the soil.

It must be appreciated, however, that no universal extraction method exists at the present time. Some guidance as to which methods work best with which groups is given in Table II, but even here such factors as sampling design and the nature of the soil must be considered.

As an example of the factors that determine the choice of sampling systems three specific projects are considered in Section V.

References

Alexander, F. E. S. and Jackson, R. M. (1955). *In* "Soil Zoology" (D. K. McE. Kevan, ed.), pp. 433–439. Butterworth, London. Preparation of sections for study of soil micro-organisms.

Anscombe, F. J. (1950). *Biometrika* **37,** 358–382. Sampling theory of the negative binomial and logarithmic series distributions.

Auerbach, S. I. and Crossley, D. A. (1960). *Acarologia* **2,** 279–288. A sampling device for soil microarthropods.

Baker, E. W. and Wharton, G. W. (1952). "An Introduction to Acarology". Macmillan, New York.

Balogh, J. (1938). *Zool. Anz.* **132,** 60–64. Vorarbeiten zu einer quantitativen Auslesemethode für die bodenbewohnenden Gliedertiere.

Balogh, J. (1958). "Lebensgemeinschaften der Landtiere", 560 pp. Akademie Verlag, Budapest and Berlin.

Belfield, W. (1956). *J. Anim. Ecol.* **25,** 275–288. The Arthropoda of a West African Pasture.

Berlese, A. (1905). *Redia* **2,** 85–89. Apparechio der raccogliere presto ed in gran numero picoli Arthropodi.

Berlese, A. (1921). *Redia* **14,** 211–214. Mezzo per separare gli artropodi raccolti col collettore Berlese de la terra caduti con essi.

Bliss, C. I. and Fisher, R. (1953). *Biometrics* **9,** 176–200. Fitting the negative binomial distribution to biological data.

Boness, M. (1953). *Z. Morph. Ökol. Tiere* **42,** 225–277. Die Fauna der Wiesen unter besonderer Berucksichtigung der Mahd.

Capstick, C. K. (1959). *J. Anim. Ecol.* **28,** 189–210. The distribution of free living nematodes in relation to salinity in the middle and upper reaches of the river Blyth estuary.

Coile, T. S. (1936). *Soil Sci.* **42,** 139–142. Soil Samplers.

Cole, L. C. (1949). *Ecology* **30,** 411–424. The measurement of interspecific association.

Clark, D. P. (1954). "The ecology of the soil fauna of a rain forest with special reference to the amphipod, *Talitrus sylvaticus* Haswell". Ph.D. Thesis. Univ. Sydney.

Cragg, J. B. (1961). *J. Ecol.* **49,** 477–506. Some aspects of the ecology of moorland animals.

Davies, M. J. (1955). "The ecology of small predatory beetles, with special reference to their competitive relations". D.Phil. Thesis. Univ. Oxford.

Dethier, V. G. (1947). "Chemical Insect Attractants and Repellents", 290 pp. McGraw Hill, London.

Dhillon, B. S. and Gibson, N. H. L. (1962). *Pedobiologia* **1,** 189–209. A study of the Acarina and Collembola of agricultural soils.

Edwards, C. A. (1958). *Ent. exp. et appl.* **1,** 308–319. The ecology of Symphyla. Part I: Populations.

Evans, G. O. (1953). *Ann. Mag. nat. Hist.,* Ser. 12. **64,** 258–281. On a collection of Acari from Kilimanjaro, Tanganyika.

Evans, G. O. (1955). *In* "Soil Zoology" (D. K. McE. Kevan, ed.), pp. 421–424. Butterworth, London. Notes on preserving, clearing, mounting and storing terrestrial mites.

Fager, E. W. (1957). *Ecology* **38,** 586–595. Determination and analysis of recurrent groups.

Finney, D. J. (1946). *Biomet. Bull.* **2**, 1–7. Field sampling for the estimation of wireworm populations.

Ford, J. (1937). *J. Anim. Ecol.* **6**, 98–111. Fluctuations in natural populations of Collembola and Acarina.

Forsslund, K. H. (1948). *Medd. Skogsforsk. Inst. Stockh.* **37** (7), 1–22. Über die Einsammlungsmethodik bei Untersuchungen der Bodenfauna.

Gisin, H. (1943). *Rev. suisse Zool.* **50**, 131–224. Ökologie und Lebensgemeinschaften der Collembolen im schweizerischen Excursionsgebiet Basels.

Gisin, H. (1960). "Collembolenfauna Europas", 312 pp. Museum d'Histoire Naturelle, Geneva.

Haarløv, N. (1947). *J. Anim. Ecol.* **16**, 115–121. A new modification of the Tullgren apparatus.

Haarløv, N. and Weis-Fogh, T. (1953). *Oikos*, **4**, 44–57. A microscopical technique for studying the undisturbed texture of soils.

Hairston, N. G. (1959). *Ecology* **40**, 404–416. Species abundance and community organization.

Hartenstein, R. (1961). *Ecology* **42**, 190–194. On the distribution of forest soil micro-arthropods and their fit to "contagions" distribution factors.

Healy, M. J. R. (In press). *In* "Research Methods in Soil Zoology" (Murphy, ed.). Butterworth, London. Some basic statistical techniques in soil zoology.

Heydemann, B. (1958). *In* "Lebensgemeinschaften der Landtiere" (J. Balogh, ed.). Budapest. Erfassungsmethoden für die Biozönosen der Kulturbiotop.

Hobart, J. (1956). *Ent. mon. Mag.* **92**, 227–228. The use of *Polyporus* for plugging small vials.

Kendall, M. G. (1955). "Rank Correlation Methods", 2nd edn., 196 pp. Charles Griffin, London.

Kevan, D. K. McE. (1955). "Soil Zoology. Proceedings of the University of Nottingham Second Easter School in Agricultural Science". Butterworth, London.

Kontkanen, P. (1950). *Vie et Milieu* **1**, 121–130. Sur les diverses methodes de groupement des recoltes dans la biocénologique animale.

Kontkanen, P. (1957). *Cold Spr. Harb. Symp. quant. Biol.* **22**, 373–378. On the delimitation of communities in research on animal biocenotics.

Kubiëna, W. L. (1938). "Micropedology". Collegiate Press, Iowa.

Kühnelt, W. (1948). *Mikroskopie* **3**, 120–128. Mikroskopie der Bodentiere.

Kühnelt, W. (1961). "Soil Biology, with special reference to the Animal Kingdom". Faber, London.

Ladell, W. R. S. (1936). *Ann. appl. Biol.* **23**, 862–879. A new apparatus for separating insects and other arthropods from the soil.

Lange, W. H., Akesson, N. B. and Carlson, E. C. (1955). *In* "Soil Zoology" (D. K. McE. Kevan, ed.), pp. 351–355. Butterworth, London. A power-driven, self-propelled soil sifter for subterranean insects.

Macfadyen, A. (1952). *J. Anim. Ecol.* **21**, 87–117. The small arthropods of a *Mollinia* fen at Cothill.

Macfadyen, A. (1953). *J. Anim. Ecol.* **22**, 65–77. Notes on methods for the extraction of small soil arthropods.

Macfadyen, A. (1955). *In* "Soil Zoology" (D. K. McE. Kevan, ed.), pp. 315–332. Butterworth, London. A comparison of methods for extracting soil arthropods.

Macfadyen, A. (1961). *J. Anim. Ecol.* **30**, 171–184. Improved funnel-type extractors for soil arthropods.

Macfadyen, A. (1962). "Animal Ecology: Aims and Methods", 2nd edn. Pitman, London.

Madge, D. S. (1961). *Nature, Lond.* **190**, 106–107. "Preferred temperatures" of land arthropods.

Morris, H. M. (1922). *Bull. ent. Res.* **13**, 197–200. On a method of separating insects and other arthropods from soil.

Murphy, P. W. (1955). *In* "Soil Zoology" (D. K. McE. Kevan, ed.), pp. 338–340. Butterworth, London. Notes on processes used in sampling, extraction and assessment of the meiofauna of heathland.

Murphy, P. W. (1958). *Ent. exp. et appl.* **1**, 94–108. The quantitative study of soil meiofauna.

Nielsen, C. Overgaard (1948). *Nat. Jutland.* **1**, 271–277. An apparatus for quantitative extraction of nematodes and rotifers from soil and moss.

Nielsen, C. Overgaard (1953). *Oikos* **4**, 187–196. Studies on Enchytraeidae. I: A technique for extracting Enchytraeidae from soil samples.

O'Connor, F. B. (1955). *Nature, Lond.* **175**, 815–816. Extraction of Enchytraeid worms from a coniferous forest soil.

O'Connor, F. B. (1957). *Oikos* **8**, 161–199. An ecological study of the Enchytraeid worm population of a coniferous forest soil.

Ostenbrinck, M. (1954). *Meded. LandbHoogesch. Gent.* **19**, 377–408. Een doelmatige methode voor het toesten van aaltjes bestrijdings meddelen in grond met *Hoplolaimus uniformis* als proefdier.

Peachey, J. E. (1959). "Studies on the Enchytraeidae of moorland soils". Ph.D. Thesis. Univ. Durham.

Quenouille, M. H. (1950). "Introductory Statistics". Butterworth, London.

Raw, F. (1955). *In* "Soil Zoology" (D. K. McE. Kevan, ed.), pp. 341–346. Butterworth, London. A flotation extraction process for soil micro-arthropods.

Salt, G. and Hollick, F. S. K. (1944). *Ann. appl. Biol.* **31**, 52–64. Studies of wireworm populations: a census of wireworms in pasture.

Satchell, J. E. (1955). *In* "Soil Zoology" (D. K. McE. Kevan, ed.), pp. 180–201. Butterworth, London. Some aspects of earthworm ecology.

Snedecor, G. W. (1946). "Statistical Methods", 4th edn. Iowa.

Strenzke, K. (1952). *Zoologica, Stuttgart.* **104**. Untersuchungen über die Tiergemeinschaften des Bodens: Die Oribatiden und ihre Synusien in der Böden Norddeutschlands.

Taylor, L. R. (1961). *Nature, Lond.* **189**, 732–735. Aggregation, variance and the mean.

Tribe, H. T. (1960). *In* "The Ecology of Soil Fungi" (D. Parkinson and J. S. Waid, eds.), pp. 246–256. Liverpool Univ. Press, Liverpool. Decomposition of buried cellulose film, with special reference to the ecology of certain soil fungi.

Tribe, H. T. (1961). *Soil Sci.* **92**, 61–77. Microbiology of cellulose decomposition in soil.

Tullgren, A. H. (1918). *Z. angew Ent.* **4**, 149–150. Ein sehr einfacher Ausleseapparat für terricole Tierfaunen.

Ulrich, A. T. (1933). *Mitt. Forstwirt. Forstwiss.* **4**, 283–323. Die Makrofauna der Waldstreu.

Van der Drift, J. (1959). *I.T.B.O.N. Meded.* **41**, 79–103. Field studies in the surface fauna of Forests.

Walker, T. J. (1957). *Ecology* **38**, 262–276. Ecological Studies of the arthropods associated with certain decaying materials in four habitats.

Waters, W. E. (1955). *Forest Sci.* **1**, 68–79. Sequential sampling in forest insect surveys.

Williams, G. (1959). *J. Anim. Ecol.* **28**, 1–14. The seasonal and diurnal activity of the fauna sampled by pitfall traps in different habitats.

Williams, T. D. and Winslow, R. D. (1955). *In* "Soil Zoology" (D. K. McE. Kevan, ed.), pp. 375–384. Butterworth, London. A synopsis of some laboratory techniques used in the quantitative recovery of cyst-forming and other nematodes from soil.

Williams, W. T. and Lambert, J. M. (1959). *J. Ecol.* **47**, 83–102. Multivariate methods in Plant Ecology. I: Association analysis in plant communities.

Williams, W. T. and Lambert, J. M. (1960). *J. Ecol.* **48**, 689–711. Multivariate methods in Plant Ecology. II: The use of an electronic computer for association analysis.

The Method of Successive Approximation in Descriptive Ecology

M. E. D. POORE

Botany Department, University of Malaya, Kuala Lumpur, Malaya

I. Introduction

Plant ecology is the science of vegetation. It has many facets, but its aim can be stated simply as the attempt to understand and explain the distribution of plants in terms of habitat and history. Those interested may look at vegetation in various ways. They may have the eyes of the plant geographer, whose main concern is with the total flora of a particular region, the distribution patterns of its species and their history; or they may have the eyes of the ecologist (*sensu stricto*) who is interested rather in the causal relationships between vegetation and habitat; or of the phytosociologist who looks at the elaborate organization of plant communities and the balance of competition and inter-dependence between species; or yet of the autecologist, primarily interested in the behaviour of single species. But, although different in outlook and aims, the subject matter of all these is vegetation and their disciplines are closely associated.

It is the common experience of plant ecologists that vegetation is not a random assemblage of individuals of many species, but that plants are associated in communities, which have a definite structure and often a regular specific composition. Only the species which are adapted to a particular habitat are selected from the available propagules of the total flora of a region, and the number of these is further restricted by considerations of space and by mutual compatibility in requirements for nutrients and light. The resulting communities vary in their degree of cohesion, which is least in transitory populations of ephemerals with single-layer structure and greatest in the tropical forest with its complex development of interrelated synusiae.

The community is one of the key concepts in the science of vegetation, and data in all branches of the subject should be treated in the community context. For example the plant geographer cannot consider the migration of taxa in isolation; for every species must migrate in a chain of suitable environments and in communities where a niche for it exists. No more can the autecologist consider the behaviour of a single species out of the context of the community, for its physiological optimum is frequently modified by competition and does not correspond with its ecological optimum.

The exact description and characterization of the community thus becomes essential to all who work in the science of vegetation; for it is important that ecological findings should be related to communities which are well described. Not to do so is tantamount to carrying out a physical experiment without stating the conditions. Many methods have been developed, for both the description and the characterization of communities, which vary from subjective and qualitative to elaborately quantitative. The choice of method is a personal matter influenced partly by the purpose of the investigation and partly by the particular view taken by the worker of the nature of vegetation. But it should be remembered that, however exact the method, it cannot describe all the attributes of even the simplest community at one moment of time. Every description is an abstraction from the available data.

The argument so far is that, to be fully useful, ecological observations must be specified in terms of the community in which they are made. Nevertheless, even if this is done, no inductive generalizations are possible without classification. Without it any observation may be only an isolated case and consequently of very limited general value. This need for classification is quite independent of the subject matter; it applies equally to any objects of which we wish to think, whether it be books, musical instruments, the shape of clouds or human emotions.

Some phenomena are easier to classify than others. This depends on how distinct is the discontinuity between one class and those nearly related to it. The classes of triangle, quadrangle, pentagon etc. can be perfectly defined and there is, by their very nature, no possibility of intermediates between them. This is a very different situation from that of colours, for example, where the common categories of yellow, red and so on are capable of every shade of gradation and cannot be separated from one another except by arbitrary dividing lines in certain physical variables such as wavelength. Yet even in phenomena such as colour some classification is essential and even the crude categories of subjective impression are invaluable and constantly used.

Vegetation is among the more difficult phenomena to classify (cf. Goodall, 1954b). The difficulty is of two kinds — in the construction of the classes and in the principles to be applied to the later grouping of the classes. Some of the divergent views which have been expressed on these topics will be discussed later.

Most working ecologists have concluded as a result of extensive field experience that, in any region, plant communities may be found which occur many times with more or less similar structure and species composition and which grow in similar habitats. These are usually communities which occupy large areas; for the area of a vegetation type is often a reliable indication of its stability. Significant and stable communities, however, may sometimes occupy small areas; and, in regions which have recently suffered from intense biotic influence or catastrophic disturbance, most of the vegetation may be labile. Not all the vegetation of a region can, therefore, be attributed to these stable communities (in extreme instances only a small proportion). Nevertheless, on the grounds of their similarity to one another, they can be combined into a class, an abstract community type, about which generalizations may be made. There are various reasons why these communities may occur repeatedly; but the logical processes of classification and generalization are independent of any possible explanation of their occurrence. I think that this is the essence of the plant sociological approach to vegetation classification. It does not attempt to produce classes into which all vegetation must be fitted but it does distinguish abstract types where there is a sufficient degree of similarity and consistency between communities to warrant the construction of classes.

Classification not only organizes and systematizes the findings of any science, but can also be used as a research tool because it throws into relief the similarities and differences between phenomena. Only broad generalizations can be made from broad classifications, but the more precise the classification, the more accurate the generalization. Reasoning such as this is implicit in most phytosociological methods, but its value seems to be little appreciated. Poore (1956) examined phytosociological methods and suggested that, by utilizing these principles they might be used for the solution of ecological problems by a process of "successive approximation", a series of inferences each closer to the truth than the preceding. This theme will be restated and elaborated in this paper.

What is meant by the methods of successive approximation? Extended illustrations of the application of the method to field problems are given by Dahl (1956) and Poore (1955c); here we may look instead at a generalized example.

Let us for a moment imagine that an ecologist is starting work in a new and unknown area. What are his problems and how can he best solve them? He is faced with a variable cover of vegetation, here forest, there grassland, marshes, dwarf shrub communities and so on. On the other hand the country has marked topographic relief, different soils and a variable intensity of grazing. His first aim is broadly to relate the variation of vegetation to that of the habitat, and then progressively to refine his knowledge so that he understands the causes underlying smaller and smaller vegetational differences.

The variety is baffling to one unfamiliar with the species and the environment, and so the ecologist tries to orientate himself, to give himself some point of reference. He chooses some community (A) and describes both it and its environment in as much detail as he is able. He chooses the community because he considers that it is typical of a much wider sample; but, because he is inexperienced in this region, his choice may be wrong. This does not matter for he will have plenty of opportunity to modify or reject his choice as the investigation continues. Having become familiar with community A, he moves to community B and describes it in the same way. This consists of different species and, let us say, occurs on a completely different soil. By comparing these two (A and B) he makes the provisional inference that the difference in vegetation between A and B is due to a difference in soil. He realizes that this may be an inaccurate or incomplete explanation, but it is consistent with his observations. After he has examined a number of these stands and found in every case the same difference in habitat, his inference becomes more sure and he can erect classes of communities and classes of habitats and generalize that community type A is distinguished from community type B by a certain consistent difference in species content and that this is correlated with a difference in habitat. Nevertheless, as he continues, he is likely to find inconsistencies. These should be carefully examined, because knowledge grows by the explanation of inconsistencies. He may find, for example, a community C in a habitat which appears identical to that of B; and by the presence of charcoal he may be able to relate this to the recent occurrence of a fire. By the continued examination of similarities and differences he gradually builds up parallel classifications of vegetation types and environmental situations and becomes able to relate the two. As his experience grows he will be able to distinguish more finely and to draw more detailed inferences. This is the "method of successive approximation". The criterion which must be used to judge the validity of the inferences made is that they should be consistent with all the facts as far as they are known. No analysis can expect to be the ultimate truth, and all hypotheses are flexible and may be modified if new and contradictory evidence comes to light.

In this way a framework of points of reference is built up. Unexplained relationships, of which there are likely to be many, due, for example, to vagaries of establishment, unknown historical factors and other "accidents", may be noted and reserved for future attention. The differences that are significant will become apparent and can be subjected to more detailed and experimental examination.

The method of successive approximation would appear to be the most economical way of obtaining a comprehensive understanding of vegetational variation. It is *par excellence* a reconnaissance method but is not limited to reconnaissance. When the main framework is determined, more detailed investigations either of the community or of the habitat will reveal finer relationships. This, not earlier, is the proper stage for the entry of more strictly quantitative methods; for only at this stage will they produce results which are commensurate to the work they involve.

If we briefly compare alternative approaches we can see what advantages the method of successive approximation offers.

The very detailed examination of one area or situation, though frequently very valuable, may easily prove to be time partially wasted, because the situation investigated is unrepresentative; or it may give negative results because, through inexperience, the wrong features of the environment have been observed.

Attempts have been made recently (Goodall, 1953ab, 1954a; Hughes and Lindley, 1955) to put extensive primary survey on a quantitative basis. These workers followed detailed quantitative investigation of the vegetation and the habitat by factor analysis or ordination. This involves much more computational work and may, in the end, be less revealing because only two or three factor complexes can be abstracted and many of the incidental variations the causes of which are plain in a detailed field investigation do not become apparent.

The detailed comparison between these different research methods will be taken up later. It is a cardinal rule, however, that the methods used should be commensurate with the scale of the problem and with the results that are to be obtained.

It should be apparent from this example that the ecologist who uses the method of successive approximation to elucidate the problems of vegetation is at the same time erecting classes and organizing his knowledge in a scheme of relationships. In fact he is classifying his data, and this classification is the most succinct and economical expression of his results.

Jevons (1913) describes the process of classification in this way: "The usual mode in which an investigator proceeds to form a classification of a new group of objects seems to consist in tentatively arranging them according to their more obvious similarities. Any two objects which present a close resemblance to each other will be joined and formed into the rudiment of a class, the definition of which will at first include all the apparent points of resemblance. Other objects as they come to our notice will gradually be assigned to those groups to which they present the greatest number of points of resemblance, and the definition of a class will often have to be altered in order to admit them."

The steps used by the descriptive ecologist are the same steps which form a part of every attempt to classify. Webb (1954) has named four to which two more, the first in the following list, have been added. These are (i) selection of the stand to be described; (ii) delimitation of the stand to be described; (iii) choice of features to be described and description of the stand; (iv) the delimitation and characterization of abstract units of vegetation from the descriptions of stands; (v) the nomenclature† of vegetation units; and (vi) the arrangement of named units in some sort of a system of classification.

Each of these steps has its own problems which are, at the same time, problems in the method of successive approximation; but before considering these, we should look into some of the general principles underlying all classifications.

II. The Classification of Vegetation

Only limited advances are possible in any science without the orderly arrangement of phenomena which is known as classification. Gilmour (1951), in a brief and admirably lucid summary of the philosophical principles of classification, has rightly said that classification is an essential prerequisite of all conceptual thought. The logical assortment of phenomena and their causes which must accompany any attempt at a serious classification gives an opportunity for the examination of all the assumptions on which any system is based.

Many attempts have been made to construct classificatory systems

† The problems of nomenclature, though real, are outside the scope of this paper and will not be discussed.

for vegetation based on a variety of principles. Braun-Blanquet (1951) has listed them conveniently as classifications based on physiognomy, chorology, ecology, syngenetics (development), and floristic affinity. There has been much argument among ecologists as to which of these most accurately represents the natural relationships of plant communities.

Gilmour (1951) and Cain (1959) have re-examined critically some of the assumptions underlying classical plant and animal taxonomy, and some of their comments and conclusions are very relevant. Gilmour, especially, has examined the logical basis of classification and has made a clear and invaluable summary of its aims and methods which I make no apology for quoting in full.

"(i) Classification is a fundamental prerequisite of all conceptual thought, whatever the subject matter of that thought. Thus, as soon as we call an object 'hot', we are creating a class of hot things and assigning the object to that class.

"(ii) The primary function of classification is to construct classes about which we can make inductive generalizations. Thus, having constructed a class of hot things, we construct a further class of 'things that hurt', and, based on experience, we can make the inductive generalization that 'hot things hurt'.

"(iii) The particular classes we construct always arise in connection with a particular purpose. The class of 'things that hurt' clearly springs from a universally important purpose to human beings — the avoidance of suffering; the class of, say, 'botanists whose fathers had beards', does not, and hence has not, so far as I know, been constructed before this moment.

"(iv) The classification which we adopt for any set of objects depends on the particular field in which we wish to make inductive generalizations. Different fields of generalization call for different classifications. Thus, if we wish to investigate the connection between female authorship and murder stories, we would make quite a different classification of books from that required for generalizations about, say, American publishers and limited editions.

"(v) Clearly some classifications are of more general use than others. Thus a classification of books based on subject matter . . . is of wider use than one based on colours of bindings. The reason for this is that the basis of the first classification, namely the contents of the book as conceived by the author, affects more of its characteristics than does the basis of the second classification, namely the colour of the book as conceived by, presumably, the publisher. In short, subject matter has a stronger influence on the total characteristics of books than has binding design. In the general theory of classification, classifications which serve a larger number of purposes, are called natural, while those serving a more limited number of purposes are termed artificial.

"(vi) It is clear from the above that there cannot be one ideal and absolute scheme of classification for any particular set of objects, instead there must always be a number of classifications, differing in their bases according to the purpose for which they have been constructed. One classification may, of course, be more natural than another in the sense defined above, and if there is one factor influencing a group of objects more powerfully than any other, then a classification based on that factor will be more natural — that is to say useful for a greater number of generalizations — than any other; but the difference between this and other classifications is one of degree only and not of kind. In the classifications of books for example, a classification based on subject matter is probably more natural than any other, but one based on authorship must run it fairly close bringing in, as it does, factors of date and style, as well as to some extent, that of subject matter."

He concludes: "Clearly the primary aim of taxonomy, on this basis, must be the construction of classes of living things about which scientifically-useful generalizations can be made. Further there must be a number of different classifications, each based on different characters and each equally valid for its own purpose." If this is true of taxonomy, it cannot but be true for vegetation classification, where there is no implication of phylogenetic descent.

In his consideration of the systematics of Darwin compared with the great pre-Darwinian systematists, Cuvier, de Candolle, Adanson and Lamarck, Cain (1959) discussed the dangers of weighting the characters which are used in classification, because the exact significance of the weighted characters is rarely understood and the assessments of the importance of particular characters certainly differs from worker to worker. He maintains that taxonomy of organisms should be based on all characters, and that "phylogenetic" weighting should be limited to those characters where there is positive evidence in its favour. For example, fossils can and should be taken into account; and characters which have been proved to be adaptive as a result of recent experimental work can be belittled.

These arguments can equally well be applied to the classification of plant communities, and lead to the conclusion that the most natural classification should be one which is based on the widest possible range of characters. In certain schemes of classification undue emphasis has been given to dominance, seral relationships, fidelity and life form, each of which taken alone may give a misleading picture of vegetational relationships. (Poore, 1955c; Braun-Blanquet, 1951; Gams, 1918 and 1941.)

Jevons (1913) pointed out another most important feature of classification: "A class must be defined by the invariable presence of certain common properties." This requirement is not met by a classification of vegetation according to faithful species. Poore (1956) has quoted an example of an association, the Fabronietum pusillae (Barkman, 1950). The published table of this association contains 14 lists and 7 faithful species; but of these lists, one contains 5 of the faithful species, one 3, two 2, eight 1 and two 0. Faithful species may be very useful in the diagnosis of an association, but they cannot be used as the sole basis of definition. In fact these communities are probably correctly placed in the Fabronietum pusillae, but the association is actually defined not by the faithful species but by other unspecified characteristics such as habitat and constant species.

The "type" system used in taxonomy may seem at first sight to contradict this principle, but, as Jevons wrote, it is only "an abbreviated mode of representing a complicated method of arrangement". The "type" is a reference point which may be returned to in the light of new evidence, or to make certain of the intentions of the author who originally described a species.

If we apply these principles to the classification of vegetation, we must recognize that there are numerous ways in which it may be classified for different purposes. A chorological classification will reveal relationships and permit generalizations which would not be possible with a classification according to life form and *vice versa*. Which of the various classifications advanced is the most natural?

I think that the majority of ecologists would agree that the generalizations most valuable to them are those which establish relationships between vegetation and its habitat, and that the most natural classification, in the sense given above, would be that which faithfully reflects these relationships. Several such have been proposed and will be discussed at a later stage. It is recognized that a classification of this kind will only partially serve the needs of the plant geographer, to whom some other will be most appropriate. Nevertheless the discussion in this paper will be confined to the various systems which try in varying degrees to reflect faithfully the relationships between vegetation and habitat.

III. The Problem of Pattern

There are two great difficulties in the classification of plant communities and in the use of descriptions to establish correlations between vegetation and habitat. One is the frequent lack of clear-cut boundaries between communities — a difficulty which will be discussed later; the other is the prevalence of complex spatial patterns in vegetation. In order to classify or to argue by successive approximation it is necessary to group together things that are equivalent. It is no use including buttons or zip-fasteners in a classification of garments, although both may be an essential part of certain articles of clothing. In the same way communities of bark epiphytes or the moss flora of an erratic boulder should not be classified with forests; nor should the chemical elements be included in a classification of igneous rocks.

Phytosociological or ecological descriptions are usually made of "uniform" communities. This is again an approximation. Absolute uniformity is foreign to vegetation; but several scales of relative uniformity can be distinguished, the data from which should not be mixed up in classification or inference. Much valuable statistical work has been carried out to investigate pattern in relatively simply communities, and there is a growing understanding of the various kinds of pattern which can occur. (For references see Greig-Smith, 1957.)

If we consider an imaginary idealized habitat which is uniform in all the physical factors of the environment in so far as they affect vascular plants, various scales of pattern, or of relative uniformity, can be distinguished.

(i) At the scale of soil micro-organisms or the exploring root tip of a vascular plant the smallest volume of soil is heterogeneous; and the least irregularities may provide micro-environments for the establishment of seeds. If we wish to examine the micro-distribution of bacteria or mites and to classify the communities in which they live, the sampling unit will clearly have to be much smaller than that required for examining the distribution of vascular plants; so will the unit of classification.

(ii) Next there is a pattern determined by the size of the individual vascular plants in the community. In simple communities this may affect the distribution of other smaller plants and animals and of parasitic organisms. But in more complex communities, such as forest, the distribution of individuals of the larger and more important species may determine the distribution of the societies of the forest floor and various societies of epiphytes. According to available statistical studies, individuals of a species rarely occur at random but are more often distributed or occur in randomly distributed contagious groups. The synusiae which are partially or totally dependent on individuals of larger species also follow this pattern of distribution.

(iii) This kind of pattern is seen not only in space but in time. Watt (1947) has described a cyclical pattern of micro-habitats caused by the life cycle of the dominant, each phase of which has its own characteristic associated flora; and this kind of pattern is now recognized as occurring widely in nature. The dominant in its various phases has a reaction on the habitat which determines its associates, but the effect on the habitat is temporary and the cycle repeats itself. The mosaic theory of regeneration put forward by Aubréville (quoted by Richards, 1952) envisages a pattern similar in kind to this, but much larger in scale, involving populations of dominants rather than individuals.

These are all patterns which can occur on an ideally uniform site. But there are also many kinds of pattern in the physico-chemical environment, particularly systematic variations in soil characteristics or micro-climate associated with pattern in the topography, which are superimposed on the pattern caused by the vegetation itself.

The plant sociologist is interested in establishing correlations between the composition of vegetation and habitat conditions. He, therefore, always tries to describe his stands from uniform sites. It is not always easy to distinguish those components of the pattern which are due to the reaction of the vegetation and are repetitive from those which are caused by underlying variation in the habitat; but some approximation can usually be made, and obviously heterogeneous stands will be detected and can be discarded at a later stage in the analysis.

Even, however, when pattern due to the environment has been excluded as far as possible, decisions must be made about ways of dealing with the pattern due to the vegetation itself. Any pattern caused by the dominant which repeats itself several times on a uniform site is usually considered by the phytosociologist as being within the permitted range of variation; but the various phases of the pattern may be sufficiently important to treat separately. It is necessary to strike a balance between making categories too broad to be useful and multiplying micro-communities to a ridiculous extent. "The answer seems to be that neither alternative is necessarily right or wrong; each case should be judged on its merits and different levels of abstraction may be appropriate for different ends. But in a floristic survey of a region some considered and consistent standpoint is desirable. The most practicable resolution of the problem seems to be as follows: if one element of a mosaic occurs in very similar or identical form as a pure community independently of the other elements with which it is usually associated, it should be described separately; similarly, if one element is associated with more than one other element to form different mosaics, it should again be treated as an independent entity" (Poore, 1955b). This also provides a practicable way of dealing with the problem of synusiae.

In any community of complex structure there are smaller groups of species which are closely related ecologically and behave to a certain extent independently of the community as a whole. Such are, for example, communities of corticolous bryophytes, herb societies of the woodland floor or the groupings of annuals and geophytes which may be found in garrigue in the Mediterranean region. These are known as synusiae. Cain and de Oliviera Castro (1959) define the synusia as "a social aggregate consisting of one or a few closely related life forms occurring together and having a similar ecology... . The total complex of synusiae that simultaneously occupy the same terrain is called a phytocoenose; this the usual concept of community".

In comparing and classifying synusiae one is comparing groups of ecologically closely related species growing in a habitat which is, for them, approximately uniform. Any change in species content of the synusia is likely, therefore, to indicate directly a change in habitat. Accordingly the classification and comparison of synusiae is likely to be profitable. Various criticisms have been levelled at treating synusiae as material for study and classification (Nordhagen, 1943; Dahl, 1956) on the grounds that the various synusiae in a phytocoenose are functionally and dynamically connected with one another. This cannot be denied; but for certain purposes the other synusiae can be treated as biotic factors of the environment and one synusia isolated for more detailed study. This approach is not arbitrary; for it is clear that the synusiae operate as independent units and may replace each other within the same phytocoenose. The principles according to which synusiae may be classified will be considered later.

Where does a synusia end and a phytocoenosis begin? This depends, according to Gams (1918), on the consistency of the interrelationships between the various parts of a complex community. He gives the following example of a resolution of this problem (Poore, 1955b): "A wood is always considered as a phytocoenose, except when the correlations between the various synusiae which form it are so close that the field and ground layers can only occur under the particular tree layer concerned." This is an excellent and rational working rule and corresponds to the treatment of mosaics given above. Both groups can only be employed after examination of a large number of stands and are accordingly subjective.

The description of a phytocoenose should be made in terms of its synusiae and these should be sampled separately. One of these synusiae is likely to be dominant in the sense that it determines the presence of the other synusiae by modifying the micro-climate or the soil. Comparison between phytocoenoses should be made principally on the basis of the dominant synusia as this reflects the general ecological characteristics of the habitat more faithfully than the subordinate synusiae. The principles of classification and comparison of phytocoenoses will be considered below.

IV. Discussion of the Steps Used in the Phyto-Sociological Examination of Vegetation

A. THE CHOICE OF STAND AND THE CONCEPT OF UNIFORMITY

It has frequently been emphasized that the problems of the synecologist begin with his choice of stand.

Whatever analysis of the community he intends to carry out and however exact the techniques he intends to apply, they will be deeply influenced by this initial choice. Much stress has been laid on the "uniformity" or "homogeneity" of the stand which should be chosen for phytosociological sampling. Most field ecologists recognize that, in any region, there are stands which are more uniform than others and their choice of stand for sampling is confined to those "more uniform" areas and especially to the communities which, after careful reconnaissance, they consider to occur frequently with more or less the same species composition. Whatever the ecological basis of this uniformity and frequency of occurrence — and there are several possibilities — the characterization of these types does provide a satisfactory and logically sound basis for an empirical classification of the vegetation. This is a subjective procedure; its justification lies in its proven usefulness and the fact that the judgements of experienced ecologists tend to coincide.

The discussion of homogeneity in Goodall (1954b) is very valuable in illustrating the approach of the quantitative ecologist. He poses the central question in the introduction to his discussion: "Plant sociology is at a disadvantage compared with the other observational sciences in that the units with which it deals are not naturally delimited. The zoologist is in no doubt where one animal ends and an individual of another species begins; and the plant systematist rarely has difficulty in detecting the spatial limits of the tissue of his systematic units; but the plant sociologist must draw artificial lines separating portions of vegetation from contiguous more or less similar vegetation before he has the wherewithal for classification."

He admits that absolute homogeneity is foreign to vegetation and suggests three ways of attempting to assess homogeneity. The first is based on the random distribution of individuals of all species in the community, a requirement which he suggests would be considered too rigorous by most plant sociologists. That it certainly would be so considered is shown by the inclusion of an index of sociability in the standard description of a relevé by the Zürich-Montpellier school; and by the fact that Nordhagen, for example, takes ten samples of uniform size in each stand to obtain an estimate of heterogeneity.

The second method is to consider the relation between the variance between sample areas and their spacing. If this variance is constant irrespective of the relative position of the samples, the vegetation may be considered homogeneous. Independence of variance and spacing would need to be demonstrated for each species present in order that the whole vegetation might be regarded as homogeneous. The third method proposed is based on the joint distribution of species. If the quantities of two species in a sample area are correlated, then the vegetation is heterogenous. This, according to Goodall, implies that the environmental factors affecting at least one of them are not uniform over the area — which may depend on both species having the same (or complementary) habitat preferences, or on one modifying the environment for the other. But this depends on the size of sample and the scale of "uniformity" required of the habitat.

These three methods all aim at establishing a degree of homogeneity within a stand. But an alternative procedure would be to assess the homogeneity of so-called uniform areas of vegetation relative to the transition areas between them.

After examining the existing data he concludes that "where homogeneity has been tested on various scales it has not been found" and suggests that this should cause heart searching among those who readily assume on the evidence of a general visual impression that the vegetation at which they are looking is homogeneous.

But before relegating all classifications of vegetation based on the subjective choice and description of stands to the dustbin, we should examine carefully exactly what these results do show us and how far they affect the actual methods by which a plant sociologist works.

The principal data on which these conclusions are based are those of Goodall (1954b) and Greig-Smith (1952). In a later review Greig-Smith (1957) discusses Goodall's results and concludes that "the samples appear to have been taken within an area which would be regarded on most subjective criteria as a single community, and the investigations refer rather to pattern . . . than to differences between communities". The investigation of Greig-Smith refers to a type of vegetation, namely tropical forest, where even the phytosociologists are not agreed about the possibility of applying the subjective criteria which they find so useful in the temperate, arctic and Mediterranean regions (Mangenot, 1958; discussion). Clearly a more useful check on the validity of these methods would be to apply an extensive analysis for homogeneity in a region where distinct claims have been made for phytosociological methods. For it is possible, indeed likely, that there are kinds of vegetation where the methods advocated by Goodall are more appropriate, but this does not necessarily invalidate results based on the subjective delimitation of stands elsewhere.

It is true that "plant sociologists anxious to proceed with the work of description and classification have been only too ready to accept their own subjective evaluation of the situation, and have in general not troubled to go beyond their general visual impression of homogeneity in the stands they are studying". Nevertheless most practising plant sociologists are well aware of the problems of pattern and their causes; they are also well aware of the limitations of the concept of "minimal area", which are so closely associated with pattern.

It is well at this stage to examine exactly how the plant sociologist works, and how the application of "successive approximation" reduces the faults of what may seem to be arbitrary procedures.

Full details of the actual field methods are given for example in Nordhagen (1928 and 1943), Poore (1955b) and Ellenberg (1956). Only a very short *précis* will be given here. After extensive reconnaissance the worker chooses stands which are as far as possible uniform in general physiognomy, ecology and species composition, as far as these may be determined by inspection. They should also be stands of communities which are typical of the region and occur frequently in it. The actual area occupied by them is less important. This is the first approximation. Many assumptions have already been made. Are they justifiable? One cannot say until the data have been examined more closely, when it will be possible to reject certain descriptions which are clearly mixtures or are atypical.

It is possible to make objective tests for "homogeneity". This was done by Dahl, for example, in certain of his communities using the distribution of the dominant as the criterion of homogeneity by comparing the distribution of numbers of shoots with the Poisson series. Nordhagen assesses homogeneity by describing ten plots of a given size in each community and examining the resulting constancy diagrams.

The "minimal area" curve is frequently used to obtain some assessment of homogeneity as well as to ensure that a "representative" portion of the community is described. A standard and simple technique for determining the "minimal area" of the community is described by Braun-Blanquet (1951), Cain and de Oliviera Castro (1959), Poore (1955a) and Ellenberg (1956). The concept has been criticized by Goodall (1952, 1954a,b), Cain and de Oliviera Castro (1959), Dahl (1956) and others. It is unlikely that any truly satisfactory rigid definition will be found for the minimal area. It is probable that it is an ideal concept to which natural communities only partially approximate, because surfaces which are large enough and are absolutely uniform in their effect on all the potential species of the community are never or very rarely found in nature. Evans and Cain (1952) suggest that some arbitrary point on the species-area curve should be chosen such as the area at which a 10% increase in area leads to a 10% increase in number of species. I agree with the analysis of Dahl when he writes: "Goodall (1952, 1954b) is sceptical of giving any more fundamental definition of the minimal area (than that of Evans and Cain). However, on inspection of many of the species-area curves published, the flattening out of the curve is very impressive and the situation may be easier in practice than in theory... . The minimal area, in spite of the theoretical difficulties involved, is a very important concept for the sociologist. The minimal area gives some idea of the size of area to be analysed to give a fairly representative picture of a stand. The trained sociologist rapidly gains an impression of the size of area to choose to give representative pictures of the vegetation types in question, whether a certain sample will yield all major constituents and whether a larger sample would yield little but a few accidental species... . It is important to select areas so large that the error involved by selecting areas of unequal size is small in relation to the variance met with when different patches of the same vegetation types are compared and when different communities are compared."

By using techniques to determine the "minimal area" the plant sociologist is attempting to set himself two standards.

(i) A criterion to judge whether his sample area falls in one community or in an ecotone or area of continuous variation. The difference between the curves obtained in these two instances is so striking that there is little doubt that the method is justified.

(ii) A method of obtaining the maximum information about the stand in the most economical way.

In doing this he is not deluded into thinking that his results are absolutely valid. Frequently after recording the "minimal area" he adds to the list all these species which occur outside the sample plot but within the stand as subjectively determined. There are always the dangers that the minimal area will not include certain rare species, or that the stand, as subjectively determined, will contain accidental species; or that its limits may have been accidentally transgressed. Anomalies of this sort generally appear during the later comparison of lists and should, if possible, be subsequently rechecked in the field. The validity of the delimitation of the stand can then be checked at two points, during the determination of the minimal area and during the subsequent comparison of descriptions.

B. DESCRIPTIVE METHODS

Once the borders of the stand have been determined, the ecologist must choose a method of description; and the range of possible choice is wide — from a species list with a subjective estimate of cover, to increasingly more exact, and exacting, quantitative techniques. The most useful data for a natural classification of plant communities are descriptions which include as much information as possible about all attributes of the community. Nevertheless, any description of a "uniform" community which includes the minimum acceptable number of facts should be usable for this purpose; although the greater the amount of information, the more valuable the description. There are difficulties introduced by the comparison of lists obtained by a variety of techniques, but these are not insuperable. The task of describing and codifying plant communities is so vast that one should be prepared to accept all accurate data of whatever provenance.

Whether he is using the descriptive method as a research tool or merely to collect the raw material for classification, whether he is using strictly objective methods or subjective, the ecologist must exercise his judgement in the choice of features to be observed, features to be rejected and the scales of measurement to be used. All his descriptions, however detailed, are abstractions from the data; all his results and hypotheses are approximations. This should not be forgotten. Objective methods have the advantage that they can be repeated and checked; they may be used to examine assumptions about the nature of the community (Cain and de Oliviera Castro, 1959), and they reduce certain kinds of bias. The objections to them are less frequently realized.

There is a brief review of the problem in Poore (1955b) in which the conclusion is drawn "that current statistical methods are inappropriate for [the description of stands for classification] and that the plant sociologist should have recourse to the most accurate methods of estimation available to him. It should of course be realized that the results he obtains will be suitable only for qualitative comparison, and that any more rigid treatment is illegitimate. The proper province of plant sociological studies should be to describe vegetation and to discover and define problems for solution by more exact methods; in addition, they will often indicate what lines of future research will prove most fruitful". I see no reason to modify this conclusion substantially, but I would not exclude data collected by statistical methods provided that, in other respects, these measure up to the standards of information required.

There are a number of descriptions of the range of information which is normally required in order that the "method of successive approximation" may be successfully applied. (Braun-Blanquet, 1951; Poore, 1955a,b; Ellenberg, 1956; and, for a very comprehensive treatment, Emberger, 1957.) Success will vary according to the extent to which these requirements are met.

Much space has been devoted in the literature to the inaccuracy of subjective estimates of cover (e.g. Hope-Simpson, 1940; Smith, 1944) and there is no doubt that criticism on this score is valid, if conclusions are drawn which are more detailed than the data justify. This is not usually the case in plant sociological literature. The published description gives, to any other worker experienced in the technique, an accurate impression of the actual vegetation, but analytical comparisons are usually confined to differences in presence or absence of species, and to large and consistent differences in abundance. Under these circumstances subjective estimation is not only justified but provides the most economical way of obtaining the necessary data.

On the other hand very little attention has been paid to the disadvantages of the more rigidly quantitative methods; these may be considered to be excessively abstract and give false security.

Most statistical methods select a certain number of features of the vegetation, number of shoots, basal area, height etc. which can readily be measured, and examine these in great detail. The results are a numerical presentation of these features abstracted from the total biology of the plant community. Details of the vitality of species, the mechanism of the plant community and its periodicity are often ignored. The more qualitative the information, the more likely it is to be omitted from any description. Yet these qualitative features are likely to be quite as important in an understanding of the place of the community in the scheme of things.

Similarly the reduction of observations to a table of figures may give a sense of psychological security which is unwarranted by the method. One might cite the example of vertical point quadrats which have been very widely used in vegetation analysis, and which have recently been shown by Warren Wilson (1960) to have a very wide range of error.

Neither of these disadvantages is inherent in statistical methods as such. Refinements of technique and fuller awareness of the total biology of the community may minimize their effects. But they should not be ignored in any assessment of the value of various techniques for the study of vegetation.

There is a clear advantage in using rapid and standardized techniques in regional vegetational studies. This facilitates the task of comparing large numbers of descriptions. Their value is clearly shown by the advanced and organized knowledge of vegetation in those countries where such techniques have been applied, as opposed to those where a preoccupation with technique has stultified the extensive application of any one method.

C. Abstract Units of Vegetation

After following these procedures the ecologist's note-book contains descriptions of those communities which he thinks are uniform, and are both typical of the region and occur frequently in it.

His next step is to arrange these communities into classes or sequences according to changes in their composition or ecology. This can be done by tabulation of the data followed by inspection (see for example Nordhagen, 1928 and 1943; Braun-Blanquet, 1951; Poore, 1955a,b,c and 1956; Dahl, 1956 and Ellenberg, 1956);† or by calculating various quantitative measures such as coefficients of community (Jaccard, 1902; Kulczyński, 1928; and Sørensen, 1948) or the "climax adaptation number" and "continuum index" of Curtis and McIntosh (1951).

These methods lead either to discrete abstract units (sociations, associations etc.) or to continuous series of communities which can be arranged along various axes of variation. Which of these two results is obtained appears to depend partly on the type of vegetation examined and partly on the preconceptions of the worker about the nature of vegetation.

The question, whether vegetation is a continuum or can be divided into discrete units, has been debated ever since Gleason formulated his "individualistic hypothesis" and seems no nearer now than then to a clear unequivocal answer. Workers who have long familiarity with vegetation and who have examined the question critically can come to opposite conclusions.

For example, Dahl writes: "The repeated occurrence in nature of vegetational discontinuities is, in the writer's opinion, the basis which makes a real classification of vegetation possible. If all vegetational change were gradual, each species reacting independently of the external factors there would be no basis of classification only of typification... . The occurrence of more or less well marked vegetational discontinuities is, in the writer's opinion, an indication that we have to deal with different communities which have to be analysed separately. If the difference is not only in the frequency of one species but can be characterized by more criteria there is a basis for referring the vegetation on either side of the discontinuity to different sociological units." At the opposite extreme Curtis (1959) maintains: "Communities . . . are not precise entities of fixed and unvarying composition, but rather are loose aggregations of species, whose make-up changes from place to place and from time to time in a more or less continuous fashion. The communities, therefore, and the entire vegetation which they compose cannot be described in the exact language of physical science but must be treated in a statistical manner as a continuous variable." Both these views gain support from the data their authors bring forward. What are the reasons for this divergence?

The results which are obtained in the final tabulation or ordination

† For a detailed description of the construction and manipulation of tables see Ellenberg (1956).

of field descriptions depend on the initial selection of stand. By including or omitting certain communities it is perfectly possible to select data which can be grouped into discrete associations or which fall into continuous series. The attitude of a particular worker to the description of phytocoenoses containing a number of different synusiae, or his conception of what constitutes a mosaic or an ecotone, or the rigidity of his definition of uniformity are all factors which easily transform the kind of result which he finally obtains.

The data of Bray and Curtis (1957) from the upland forests of Wisconsin provide a useful demonstration of the way in which vegetation can be treated as a continuum, and the value in certain circumstances of analysing it as such; but it in no way proves that this vegetation is a continuum. A more detailed discussion of these results may illustrate points in this argument.

The apparently continuous variation of vegetation is probably enhanced by the method of sampling used.

(i) The areas which were used as samples were all more than fifteen acres in extent and, where possible, more than forty acres. It is extremely unlikely that such large surfaces would meet the exacting demands for site uniformity made by Dahl in his work. If even a small range of variation were included in each sample, these might overlap and the series might well show continuous variation.

(ii) Fourteen of the twenty-six species chosen for the ordination were herbs and shrubs whose occurrence is likely to be governed by site factors of a different amplitude to those of the tree synusia. This problem has been discussed above.

The first and principal axis of variation extracted from the data (the x axis) shows "mainly the relationship of the community to major past disturbance factors and to its own developmental recovery from those factors". It is probable that species composition and importance does change continuously during a succession; though there is still doubt whether some successional stages may not be passed through more rapidly than others, thus introducing discontinuities into the sere. By contrast the situation which Dahl investigated is one in which the vegetation shows distinct zonation and in which no succession of importance can take place between the different zones.

If one considers the arrangement of stands along this x axis, some stands certainly appear all along the axis; but the majority of stands are conspicuously grouped. This hardly represents a continuous distribution.

Thus in spite of the method of sampling, which would be likely to iron out discontinuities, the data of Bray and Curtis do not provide very convincing evidence for the continuous nature of vegetation if the relative number of stands occurring at various points on their axis is taken into account.

But even were their data convincing, there would be no reason to assume, therefore, that all vegetation consists of continua, or even that the upland forest communities of southern Wisconsin, if analysed in a different way, might not have presented more marked discontinuities (cf. Daubenmire, 1960).

The studies of Poore (1955a,b,c, 1956) in Breadalbane, Scotland made use of subjective methods, but special attention was devoted to the problems arising from mosaics, heterogeneity of the stand and ecotones. In the final tabulation of stands the following categories were omitted. (i) Communities which were obviously unstable; (ii) small fragments; (iii) ecotones; (iv) mosaics.

It was found that the remainder could be classified without difficulty into abstract units (noda), with the exception of certain communities which were unique in the region and others which appeared to fall into floristic series and not into discrete groups. The final conclusion from this study was: "It may well be that further research will show that most, if not all, communities fall into series, and that the discontinuities found are artifacts. But I have found that it is convenient, even in a continuous series, to establish arbitrary noda as points of reference. On the other hand, it is logically preferable, until discontinuities are actually established to assume that all communities can be arranged in series, rather than the reverse attitude which is held by the Braun-Blanquet school, whose followers maintain the distinctness of the associations as a doctrine only by the arbitrary rejection of descriptive material."

It must remain at present an open question whether or not the variation in vegetation is continuous, and it is hard to see how sufficiently extensive data could be obtained to establish either alternative as a general proposition. For such a study to prove its point it would not only have to establish that communities can be found that form continua along a number of axes of variation, but that all these communities occupy significantly comparable areas. I do not think that it is either necessary or important to find an absolute answer to this question for the result would alter very little the fundamental logical processes associated with the classification of vegetation, and would add little to our knowledge of the functioning of vegetation.

As we have seen this problem of continuity *versus* discontinuity arises at two separate and distinct stages: in defining the stand or concrete unit, and in defining the nodum or abstract unit. Discontinuities can be found between stands of vegetation due, for example, to sudden differences of soil, change of dominant species or to different histories; but continuous gradations or ecotones also occur. Similarly, in the field of abstract units, discontinuities may be detectable due to the absence of intermediate habitats; or communities may fall in continuous series. Perhaps we might use two models. In both of them individual stands of vegetation are represented by points in multidimensional space, each dimension representing a line of variation in vegetation. If we consider vegetation as essentially continuous the points will be distributed at random throughout the matrix, and no grouping but an arbitrary grouping will be possible. On the other hand, if the variation of vegetation is not continuous the points will be found to be grouped in nebulae.

If the second alternative is true it should be possible to extract and define these nebulae and to use them as points or areas of reference in terms of which "intermediate" communities may be defined. I have applied the term "nodum" to each of these nebulae (Poore, 1955b).

If the first alternative, that there is no grouping of points, is a situation which is frequently encountered, the problem of classification must be tackled somewhat differently. There are three possibilities. (i) Certain arbitrarily chosen stands should be used as types; (ii) stands should be defined with reference to their position by a series of coordinates; (iii) the matrix should be divided arbitrarily into a series of blocks defined by the sum of the characters of the stands within the blocks.

Any results may then be referred to a named vegetation unit or defined area of study — they are not left floating in a multidimensional void.

Nevertheless a frame of reference must not be allowed to become a prison cell. Concepts that have led to great advances in science have sometimes by their very success, become sacrosanct. They have in their later stages discouraged a wider vision and have actually impeded progress. Certain definitions of the association and the climax, for example, though leading to immediate advances in understanding have later left their exponents in mental blinkers and have restricted rather than assisted their work. The notions of "nodum" and of the potential continuous variation of vegetation are flexible concepts which should not operate restrictively.

In previous papers (Poore, 1955a,b,c, 1956) I have distinguished noda, because it seemed clear that, in the areas examined, the communities could be grouped into abstract units without distortion of the facts. But there would be value in the concepts of "type" or "nodum" even if it were to be established that the variation of vegetation was continuous; although the noda would then lose all their absolute validity and become arbitrary groupings.

It is extremely difficult, even in limited studies, to grasp directions of variation without reference to particular fixed points. Moreover, in extensive treatments of vegetation the data become incomprehensible unless they can be arranged in classes. These two propositions may be illustrated by the works respectively of Bray and Curtis (1957) and Curtis (1959). In the former the axes of variation are constructed between reference points. In the construction of the x axis, for example, these consisted of two sets of three stands each. These two sets were widely separated from each other, but the index values of the three stands within each set were highly significantly correlated. These sets correspond exactly to the concept of "noda" and have been used by these authors as points of reference. In the latter work, which is an extensive treatment of the vegetation of Wisconsin, arbitrary classes are erected, "Southern Forests", "Northern Forests", "Prairie", "Sand barrens and bracken grassland" etc., and these are sub-divided into, for example, "Southern Forests mesic", "Southern Forests xeric" and "Southern Forests lowland".

This erection of points of reference or of classes cannot be avoided if the data are to be presented in a comprehensible and useful form, irrespective of whether the classes are arbitrary or natural. The nature of variation within and between these classes may then be analysed as the individual worker wishes, by ordination, by factor analysis or by "successive approximation".

V. Systems of Classification and Criteria for Defining Noda

A number of different systems of classification have been proposed which variously illuminate the relationships between vegetation and habitat, and thus fulfil the requirements which were earlier suggested for a natural classification. These fall into two principal groups, the physiognomic-ecological and the floristic; but the distinctions between them are frequently blurred and they often lead to the same major units.

There is not only the question of the principles of classification to be considered, but also the status of the units to be classified. It was argued above that synusiae and phytocoenoses are different in kind and cannot be classified together. In some instances the examination and classification of synusiae can have great value; indeed it is the only practicable way of studying dependent societies and results of importance have been obtained by it, as, for example in the study of epiphytic bryophyte communities in Japan (Hosokawa, 1956; Omura *et al.*, 1955; Omura and Hosokawa, 1959). These cannot be classified on the same level as the forests in which they occur; neither can all the small changes which occur in the make-up of the dependent societies be taken into account in a classification of the forest. The confusion which arises when an attempt is made to classify phytocoenoses using all the constituent synusiae is illustrated by Cain and de Oliviera Castro (1959) from data from the red spruce association (Piceetum rubentis) of the Great Smoky Mountains in which a possible total of 256 separate phytocoenoses might occur, all with the same superior arborescent layer but with various combinations of layer-societies in the four lower strata. It may sometimes be convenient to study the phytocoenose, sometimes the synusia, but the principles which are applied to the classification of one may not necessarily apply to the other.

Some systems of classification have arisen in the attempt to produce a world-wide or continental conspectus of vegetation; such for example are the works of Schimper (1898), Schimper and von Faber (1935), Weaver and Clements (1938) and Rübel (1930). Others have arisen by generalizing from more local studies, as has that of Braun-Blanquet (1951). That these do not altogether coincide in their higher units may be due to the fact that the same methods are not appropriate for extensive and more local analysis. The highest unit of the Braun-Blanquet hierarchy, being, as it is, floristically based, is the Community circle (Gesellschaftskreis) which is essentially a plant geographical unit. This is defined as: "The final unit is a system of vegetation based on floristics. It is characterized by the total complement of all species and communities which are peculiar to or favour one natural Gesellschaftskreis (Vegetation region) and derives its most comprehensive value through an independence which is partially based on climate and partially on floristic history." Similarly in his approach to the classification of Malaysia, van Steenis (1950) uses the same concept. But the higher categories of Schimper and Rübel transcend these floristic barriers.

A. Biocoenose and Geobiocoenose

Attention is being increasingly drawn to biocoenoses as a subject for study. Sukachev (1944) has extended this concept to include the physical elements associated with living organisms in any one site. Only in the geobiocoenoses can the interrelationships of rock, soil, plants and animals in all their complexity and the energy relationships between them be studied.

There are real difficulties at present in the way of a classification of biocoenoses. This is partly due to the divergent developments of plant and animal ecology and partly to problems raised by the behaviour of the animal component, many of which move daily or at particular seasons from one biocoenose to another and must properly be considered a part of both. To what biocoenose, for example, do migratory flocks of geese belong or swarms of the desert locust? It is by no means clear on what principles a classification of biocoenoses should develop. Sukachev proposes one based on "development and the resultant complexity of sociologic structure". He maintains that a natural classification of phytocoenoses must rest upon features which are involved in transformations of matter or energy.

At present it would seem most reasonable to classify the biocoenose by means of the dominant phytocoenose, using animals where necessary, as part of the characterization, and treating those which migrate from one phytocoenose to another as transgressive elements.

B. PHYLOGENY OF PLANT COMMUNITIES

As the study of phylogeny has so much influenced taxonomy, it may be useful to look into its possible value in the study of vegetation. The very great similarity that can sometimes be found between communities which are far apart from one another is strong *prima facie* evidence for the historical continuity of these communities. Table I compares the

TABLE I

Comparison of Degree of Presence of Certain Species in Calliergon sarmentosum*—rich* Carex saxatilis *Association (Nordhagen) and* Carex saxatilis *Sociation (Poore)*

	Presence per cent	
	NORWAY	SCOTLAND
Selaginella selaginoides	0	70
Carex demissa	0	70
C. dioica	0	60
C. nigra	50	80
C. saxatilis	100	100
Eriophorum angustifolium	95	90
Festuca vivipara	0	60
Juncus triglumis	20	70
Polygonum viviparum	50	80
Acrocladium sarmentosum	100	30
A. stramineum	60	20
Aneura pinguis	0	10
Drepanocladus revolvens	100	100
Hylocomium splendens	0	100
Polytrichum commune	0	80
Scapania undulata	10	100
Scorpidium scorpioides	90	40

Only species with a degree of presence of 60% or more in either of the noda are included. The following additional species were common to the two noda: *Juncus biglumis, Deschampsia caespitosa, Leontodon autumnalis, Pinguicula vulgaris, Saxifraga stellaris, Salix herbacea, Acrocladium trifarium, Campylium stellatum, Drepanocladus exannulatus, Cinclidium stygium, Tayloria lingulata, Scapania uliginosa.*

Nomenclature in the Table is that used in Poore (1955c).

Carex saxatilis sociation found by Poore in Central Scotland (Poore, 1955c) with the Scandinavian sociation for Sylene in Central Norway (Nordhagen, 1928). It seems very probable that these are local segregates of a post-glacial sociation from which the present representatives have originated by subsequent divergent migration.

Fossil evidence supports the occurrence of such migration of the plant community as a structural unit. Table II is taken from Cain (1944) and compares the common list of trees from the modern forest at Jasper Ridge, Palo Alto, California, and the matching species of the Miocene Mascall Flora (Chaney, 1925).

TABLE II

A Comparison of the Common Trees of the Modern Forest at Jasper Ridge, Palo Alto, California, and the Matching Species of the Miocene Mascall Flora

Jasper Ridge Flora	*Mascall Flora*
Sequoia sempervirens	*Sequoia langsdorfii*
Salix laevigata	*Salix varians*
Salix lasiolepis	*Salix varians*
Populus trichocarpa	*Populus lindgreni*
Alnus rhombifolia	*Alnus* sp.
Quercus Kelloggii	*Quercus pseudo-lyrata*
Quercus agrifolia	*Quercus convexa*
Quercus lobata	*Quercus duriuscula*
Umbellularia californica	*Umbellularia* sp.
Arbutus Menziesii	*Arbutus* sp.
Acer macrophyllum	*Acer bolanderi*
Aesculus californica	*Absent*

From Cain, 1944

In the course of time communities migrate in response to secular changes of climate; and they change as they migrate. The evolution of new species contributes to this change, but it is also brought about by the break up and extinction of parts of the community and by segregation or recombination of its elements.

Evidence is found in the post-glacial history of British vegetation of the segregation of grassland and ruderal communities from species which occupied a rare and sporadic place in the intervening woodland phase (Godwin, 1956).

In a discussion of Leskov's work on an *Abies* forest Sukachev (1944) shows how in his view *Pinus* forests have not differentiated out of one primitive stock but have arisen though the Pine colonizing and combining with other and different floristic elements. This does not only apply to the dominants. Of eight species co-dominant on the floor of the Pinetum oxalidosum four (*Asarum europaeum*, *Asperula odorata*, *Maianthemum bifolium* and *Trientalis europaea*) probably developed in the Tertiary under deciduous forest with dense summer shade, while four (*Lycopodium annotinum*, *Vaccinium myrtillus*, *Pyrola secunda* and *Linnaea borealis*) are related to coniferous forests.

It is possible that ecological groups of species, rather similar to synusiae, may have had a common origin and have operated as units throughout vegetation history but even this is doubtful and certainly unproven.

Although historical evidence supports the existence of certain communities relatively unchanged through long periods of geological time, it shows that this is by no means invariable. The history of many plant communities is complicated and major associations of the present day may be drawn from elements in many different past communities. Phylogeny proves therefore to be a very unsatisfactory and uncertain basis for the classification of plant communities, although anyone who studies the history and development of vegetation may draw very valuable comparative data from a satisfactory classification of present-day plant communities.

Having rejected the phylogeny of plant communities as the basis for classification, the problem is to find means of comparing and arranging phytocoenoses on the one hand and synusiae on the other into systems that will throw into relief the relationship between their composition and structure and the habitat.

C. DEFINITION OF COMMUNITIES

In the discussion of classification given above it was suggested that a natural classification should be based on all available characteristics, and that a class should be defined by the "invariable presence of certain common properties". If we want to distinguish vegetation units for later correlation with habitat or with animal distribution, it is necessary that the units should be distinguished by characteristics of the vegetation. No features of the habitat should be used in the definition. If, however, we want to define vegetation-habitat complexes no such restriction is necessary.

D. PHYSIOGNOMIC-ECOLOGICAL OR FLORISTIC CLASSIFICATION?

The purpose of a natural classification of vegetation is to obtain a definition and arrangement of plant communities which reflect as closely as possible their relationship to one another as determined by their total response to their environment. Communities which are ecologically similar should be grouped together.

Within a small area this is easy, for floristic composition and ecology usually alter together and consistently with one another. The change in the Scottish mountains from a *Rhacomitrium-Carex bigelowii* sociation to one of *Rhacomitrium* and *Juncus trifidus* is regularly correlated with a particular shift in environment. The whole floristic composition usually reflects finer differences than does either the dominant or the structure of the community. Moreover the physiognomy can be reconstructed from a good description in floristic terms, but floristics cannot be inferred from a description of structure.

On a global scale, however, this method does not apply because areas which appear to be ecologically equivalent have different floristic histories. A truly ecological classification would unite the equatorial forests of Malaysia, Central Africa and Tropical America more closely than any one of them to the sub-tropical savannahs adjoining it. A physiognomic classification would do the same because of the striking homoplastic evolution of vegetation structure in response to habitat. A classification based on floristics would, on the other hand, separate them widely.

These considerations lead to the conclusion that classification of communities within one relatively uniform floristic region should be based on floristic comparison, but that classification on a larger scale should be based on physiognomic-ecological criteria. For this the system proposed by Gams in 1918 supplemented by the decisions of the International Botanical Congress (Du Rietz, 1936) still seem to be the most satisfactory.

Taken together these recommend that there should be separate classifications of synusiae and phytocoenoses. The units of these two systems, within a uniform floristic region, being as follows.

Synusiae	*Phyto- (bio-) coenoses*
Society	Sociation
Union	Association
Federation	Alliance

When this system is extended outside a uniform floristic region, the following concepts and terms are appropriate.

For synusiae: isoecia; abstract units built up of ecologically equivalent but floristically different synusiae of different regions.

For phytocoenoses: isocoenoses; abstract units built up of floristically different but ecologically indentical consociations.

E. Definitions of Units

Definitions of units should be precise so that there is no doubt whether or not a community falls into a named unit. If we accept the arguments developed above, however, the criteria used will be different in small and in large regions. Much of the confusion which has arisen in the classification of vegetation can be attributed to attempts to apply the same criteria to define all units.

Isoecia and isocoenoses should be distinguished by their physiognomy and ecology, the remainder of the units by their floristic composition. The floristic criteria available are the species occurring in a community and their dominance, constancy and fidelity. Braun-Blanquet (1951) and Poore (1955c) have given objections to the sole use of dominance, Gams (1941) and Poore (1955c) to fidelity.

Constancy appears the most unexceptionable criterion, because a definition based on constancy limits the range of the "nodum" to that in which the ecological amplitudes of those species named as constants overlap. I would suggest as the definition of a "union" or "association" the following: "A nodum defined in terms of its constant species" and any stand would be attributed to that unit in which more than 80% of the species named as constant to the unit were to be found.

In the synusial classification this definition would require no qualification: in the phytocoenosal classification the condition of constancy should refer only to the dominant synusia.

F. MULTIDIMENSIONAL VARIATION

Many authors have concluded that variation in vegetation is multidimensional and that the most satisfactory way of representing vegetational variation in a region is by diagrams illustrating lines of variation (cf. Poore, 1955c; Gams, 1941; Agnew in Greig-Smith, 1957). Units can alternatively be separated by keys; and it should be possible to construct two parallel keys in any region, one of vegetation and one of habitat which coincide with one another. Curtis in *The Vegetation of Wisconsin* has constructed keys; but habitat and vegetation are used as diagnostic characters in the same key.

Although the relationships of vegetation are undoubtedly multidimensional and it may be very useful to arrange synusiae and phytocoenoses in diagrams to illustrate relationships of various kinds, some sort of hierarchical arrangment is necessary to make the available data manageable and comprehensible. Several arrangements of this kind are possible, some of which are more generally useful than others.

The system of Braun-Blanquet is unsatisfactory at the higher levels because it limits the hierarchy to one Vegetationskreis, and therefore widely separates communities of equivalent ecology. Van Steenis (1950) also limits his scheme to one vegetation area but makes his subordinate classification according to a hierarchy of habitat factors. He is classifying vegetation by a criterion that is not vegetational.

Clements and Tansley classify according to the climax or climax mosaic, and thereby combine in one hierarchy communities which are poles apart ecologically, and separate those which are ecologically similar.

The physiognomic-ecological classification seems to be the most natural for our purpose, as the physiognomy and structure of the phytocoenose seem to reflect rather faithfully the sum total of the ecological factors of the habitat. I would suggest therefore that the major units of a vegetational classification should be physiognomic, while the lower units may be arranged in floristic series. It is perhaps notable that this is the treatment which Curtis has followed in his *Vegetation of Wisconsin.* I would like also to suggest some tentative additional reasons why physiognomic units may be the most satisfactory.

It is very striking that similar structural patterns occur under similar environments in different parts of the world, even when there are great geographical discontinuities between them and the floras are very different. When the same structural pattern extends over large areas of country it is noteworthy that there is a gradual change in the floristic composition of the community without a change in structure. Lippmaa (1939) has worked this out in detail for the *Galeobdolon-Asperula-Asarum* union of the Baltic deciduous forest region; Poore and McVean (1957) describe such series across Scotland from the oceanic west to the more continental east — series which continue into Scandinavia. The place occupied by one species in the structure of one community is exactly filled by an equivalent species of the same life form in another. The two species may be of the same genus, but may be of related genera if the distance between the communities is great.

This suggests that there are only a limited number of structural patterns which occur in vegetation, and that these have considerable inherent stability. It is certainly common experience that ecotones between communities that are structurally different are much more abrupt than between structurally identical communities. Also, although mosaics may occur composed of communities which have a different physiognomy, true intermediates rarely occur. Those stages of a succession, too, which involve a total change of structure, are usually relatively rapid.

VI. Discussion

The structure of vegetation is very complex and its variation in space and time is determined by a multiplicity of factors or factor complexes whose effects and interactions are often little understood. Hence experiment, the normal tool of the scientist, is of limited usefulness.

It is true that some vegetational situations can be submitted to experiment, especially, perhaps, those concerned with the operation of biotic factors such as grazing and fire; and, wherever experimental synecological methods can be used to test hypotheses, this is desirable. Experimental methods may also be applied to autecological problems; but, such is the operation of competition, that the results of controlled culture are often of very limited application to the situation in the field.

The ecologist should not ignore any avenue open to experiment and he should be fully aware of the implications to the phenomena which he is studying of the work of the experimental biologist. Nevertheless he must frequently use other means of investigation and reasoning to try to elucidate the problems that face him.

The best of these is extensive, critical observation coupled with the classification of data. By following this method correlations can be discovered, hypotheses formed, and checked and rechecked for consistency by further observation. Any inconsistency leads to the reformulation of the hypotheses. The hypothesis held at any time is that which is consistent with all the data available. No hypothesis can be more. It is this well-known part of scientific method that I have called "successive approximation". Its principles are the same as those of classification, and the method, if applied extensively, leads inevitably to a classification of the phenomena being studied. The method has frequently been used for reconnaissance, but need not be confined to superficial observation. The detail of the results will be commensurate with the detail of observations.

Cain (1944) wrote: "The complex interrelations among the environmental factors, and between them and the organism, with its complex physiological and morphological interrelations, are such as to defy solution in exact terms of causation. Ecological problems may not only be difficult of solution because of the interaction of factors and responses, but they may really be insoluble in a mathematical sense."

It is very important therefore that any detailed situation which is investigated by experimental methods or by time-consuming quantitative methods should be representative of a number of situations and such that useful generalizations may be drawn from it.

The alternative, rigidly quantitative approach of statistical description and factor analysis cannot, by the sheer volume of work it necessitates, provide the wide coverage necessary for the initial assessment of critical problems. By choosing only those features which are quantitative, it abstracts from the full biology of the community, and the analysis, in those instances when it has been fully applied, leads to the extraction of environmental complexes which are already known. Applied, however, to critical situations of the detailed distribution of species and to small environmental differences, it may well obtain results commensurate with effort.

Part of the dilemma of descriptive ecology stems from the late development of the science. In taxonomy the great volume of descriptive work had already been accomplished before the advent of statistical treatments. A framework of major groupings had already been built within which biometric methods could be used to the fullest advantage in checking earlier judgements and the elucidation of difficult problems. If taxonomy were now in the state of descriptive ecology it is likely that there would be a questioning of the whole base of descriptive taxonomy and that it would be argued that no description of species should be published unless preceded by a thorough biometric analysis of populations and the establishment of demonstrated discontinuities between taxa. The effect on orthodox taxonomy would be disastrous, and would render impossible the various sciences, such as ecology, which depend for their existence on the findings of taxonomy, however approximate and imperfect these may be.

Acknowledgement

I wish to acknowledge the assistance of L. F. H. Merton who has helped me to clarify these ideas in many long discussions and who has criticized this paper in manuscript.

References

Barkman, J. J. (1950). Synopsis of address to the *Int. Bot. Congr.*, Stockholm.

Braun-Blanquet, J. (1951). "Pflanzensoziologie", 2nd edn. Springer Verlag, Vienna.

Bray, J. R. and Curtis, J. T. (1957). *Ecol. Monogr.* **27**(4), 325–349. An Ordination of the Upland Forest Communities of Southern Wisconsin.

Cain, A. J. (1959). *Proc. Linn. Soc. Lond.* **170**(2), 185–217. Deductive and inductive methods in post Linnaean taxonomy.

Cain, S. A. (1944). "Foundations of Plant Geography". Harper, New York and London.

Cain, S. A. and de Oliviera Castro, G. M. (1959). "Manual of Vegetation Analysis". Harper, New York.

Chaney, R. W. (1925). *Publ. Carneg. Instn.* **349**, 23–48. The Mascall Flora — Its distribution and climatic relation.

Curtis, J. T. (1959). "The Vegetation of Wisconsin. An Ordination of Plant Communities". Madison.

Curtis, J. T. and McIntosh, R. P. (1951). *Ecology* **32**, 476–496. An upland forest continuum in the prairie-forest border region of Wisconsin.

Dahl, Eilif (1956). *Skr. norske VidenskAkad.* **I.** *Nat.-Naturv.* Kl. 3. Rondane; mountain vegetation in south Norway and its relation to the environment.

Daubenmire, R. (1960). *Silva fenn.* **105**, 22–25. Some major problems in vegetation classification.

Du Rietz, G. E. (1936). *Svensk bot. Tidskr.* **30**, 580. Classification and nomenclature of vegetation units. 1930–1935.

Ellenberg, H. (1956). "Einführung in die Phytologie" von Heinrich Walter. IV. Grundlagen der Vegetations-gliederung. I Teil. Stuttgart.

Emberger, L. (1957). *Bull. Serv. Carte Phytogeographique,* Serie B. **2**(2), 7–35. Description et mode d'emploi d'une fiche de relevé pour l'inventaire de la végétation.

Evans, F. C. and Cain, S. A. (1952). *Contr. Lab. Vertebr. Biol. Univ. Mich.* **51**, 1–17. Preliminary studies on the vegetation of an old-field community in south eastern Michigan.

Gams, H. (1918). *Vjschr. naturf. Ges. Zürich* **43**. Principienfragen der Vegetationsforschung.

Gams, H. (1941). *Bot. Arch.* **42**, 201. Über neue Beiträge zur Vegetationssystematik unter besonderer Berücksichtigung des floristischen Systems von Braun-Blanquet.

Gilmour, J. S. L. (1951). *Nature, Lond.* **168**, 400–402. The development of taxonomic theory since 1851.

Godwin, H. (1956). "The History of the British Flora". Cambridge Univ. Press.

Goodall, D. W. (1952). *Biol. Rev.* **27**, 194–245. Quantitative aspects of plant distribution.

Goodall, D. W. (1953a). *Aust. J. Bot.* **1**, 39–63. Objective methods for the classification of vegetation. I. The use of positive interspecific correlation.

Goodall, D. W. (1953b). *Aust. J. Bot.* **1**, 434–456. Objective methods for the classification of vegetation. II. Fidelity and indicator value.

Goodall, D. W. (1954a). *Aust. J. Bot.* **2**, 304–324. Objective methods for the classification of vegetation. III. An essay in the use of factor analysis.

Goodall, D. W. (1954b). *Festschrift Aichinger* **1**, 168–182. Vegetational Classification and Vegetation continua. Angewandte Pflanzensoziologie.

Greig-Smith, P. (1952). *J. Ecol.* **40**, 316–330. Ecological observations on degraded and secondary forest in Trinidad, British West Indies. II. Structure of the communities.

Greig-Smith, P. (1957). "Quantitative Plant Ecology". Butterworth, London.

Hope-Simpson, J. F. (1940). *J. Ecol.* **28**, 193–209. On the errors in the ordinary use of subjective frequency estimations in grassland.

Hosokawa, T. (1956). *Jap. J. Ecol.* **5**, 93–100 and 150–3. On the structure of the Beech Forest of Mt. Hiko, S.W. Japan.

Hughes, R. E. and Lindley, D. V. (1955). *Nature, Lond.* **175**, 806–807. Application of biometric methods to problems of classification in ecology.

Jaccard, Paul (1902). *Flora*, 349–377. Gesetze der Pflanzenverteilung der Alpinen Region.

Jevons, W. S. (1913). "The Principles of Science". Macmillan, London.

Kulczyński, St. (1928). *Bull. int. Acad. pol. Sci.* (1927) No. Supp. II, 57–203. Die Pflanzenassoziationen der Pieninen.

Lippmaa, Y. (1939). *Amer. Midl. Nat.* **21**. The unistratal concept of plant communities (the unions).

Mangenot, G. (1958). *Proceedings of the Kandy Symposium U.N.E.S.C.O. Humid Tropics Research*, 115–126. Les recherches sur le végétation dans les régions tropicales humide d'Afrique occidentale. Study of Tropical Vegetation.

Nordhagen, R. (1928). *Norske Vidensk. Akad. Skrift.* Mat: naturw. Klasse, 1 (1927), 612 pp. Oslo. Die Vegetation und Flora des Sylenegebiets.

Nordhagen, R. (1943). *Bergens Mus. Skr.* **22**. Sikilsdalen og Norges Fjellbeiter.

Omura, M. and Hosokawa, T. (1959). *Memoirs Fac. Sci. Kyushu University*, Series E. **3**(1), 51–63. On the detailed structure of a corticolous community on the basis of interspecific association.

Omura, M., Nishihara, Y. and Hosokawa, T. (1955). *Revue bryol. lichen.* **24**, 59–68. On the Epiphyte Communities in Beech Forests of Mt. Hiko in Japan.

Poore, M. E. D. (1955a). *J. Ecol.* **43**(1), 226–244. The use of phytosociological methods in ecological investigations. I. The Braun-Blanquet system.

Poore, M. E. D. (1955b). *J. Ecol.* **43**(1), 245–269. The use of phytosociological methods in ecological investigations. II. Practical issues involved in an attempt to apply the Braun-Blanquet system.

Poore, M. E. D. (1955c). *J. Ecol.* **43**(2), 606–651. The use of phytosociological methods in ecological investigations. III. Practical application.

Poore, M. E. D. (1956). *J. Ecol.* **44**(1), 28–50. The use of phytosociological methods in ecological investigations. IV. General discussion of phytosociological problems.

Poore, M. E. D. and McVean, D. N. (1957). *J. Ecol.* **45**, 401–439. A new approach to Scottish mountain vegetation.

Richards, P. W. (1952). "The Tropical Rain Forest". Cambridge Univ. Press.

Rübel, E. (1930). "Pflanzengesellschaften der Erde". Bern-Berlin.

Schimper, A. F. W. (1898). "Pflanzengeographie auf physiologischer Grundlage". (Trans. P. Groom and I. B. Balfour as "Plant Geography Upon a Physiological Basis", Oxford, 1903.)

Schimper, A. F. W. and von Faber, F. C. (1935). "Pflanzengeographie auf physiologischer Grundlage". Jena; 3rd edn (1960), Codicote, England.

Smith, A. D. (1944). *Ecology* **25**, 441–448. A study of the reliability of range vegetation estimates.

Sørensen, Thorvald (1948). *K. danske vidensk. Selsk., Biol. skr.* **5**(4), 34 pp. A method of establishing groups of equal amplitude in plant sociology based on similarity of species content.

Sukachev, Y. N. (1944). *Zhür. Obschei. Biöl.* **5**, 213–227. Trans. Raney, F. and Daubenmire, R. (1958), *Ecology* **39**(2), 364–367.

Van Steenis, C. G. G. J. (1950). *Proc. 7th Int. Bot. Cong.*, Stockholm. 637–644. On the hierarchy of environmental factors in plant ecology.

Warren Wilson, J. (1960). *New Phytol.* **59**(1), 1–8. Inclined point quadrats.

Watt, A. S. (1947). *J. Ecol.* **35**, 1. Pattern and process in the plant community.

Weaver, J. E. and Clements, F. E. (1938). "Plant Ecology". McGraw Hill, New York and London.

Webb, D. A. (1954). *Bot. Tidsskr.* **51**, 362–370. Is a classification of plant communities either possible or desirable?

Energy in Animal Ecology

L. B. SLOBODKIN

Department of Zoology,
University of Michigan, Ann Arbor, Michigan, U.S.A.

I. The Relevance of Energy Studies to Ecology

No single measurement is intrinsically significant. All measurements derive their interest from their context and the richness of predictive generalizations that can be produced from them. The fact that energy measurements have high intellectual prestige in chemistry and physics does not necessarily imply major ecological significance, any more than electron spin measurements as such have ecological significance.

Precise measurement of energetic parameters is almost impossible, and even rather crude measurements are time-consuming and expensive. It must, therefore, be initially established that such measurements are of sufficient ecological significance to be worth our trouble.

One demonstration of the relevance and interest of energetics to ecology has been presented by Hairston *et al.* (1960). Energy-rich organic sediments accumulate at a rate that is completely insignificant compared with the rate of energy fixation by green plants. This implies that the biosphere as a whole is energy limited, although individual decomposer populations may be temporarily limited by predators or other factors.

In some situations, particularly the free water of lakes and oceans, plants are depleted by herbivores so that the herbivores are obviously limited by energy. Occasionally, it can be directly demonstrated that herbivore population-size is dependent on the rate of energy fixation by food plants (Borecky, 1956). In terrestrial situations, living vegetation is not usually depleted by the activities of herbivores. Situations in which terrestrial vegetation is eliminated often involve exotic herbivores. The success of exotic herbivores at plant destruction implies that climate does not usually regulate the size of herbivore populations, since the exotic herbivore cannot be expected to show higher adaptation to the weather than the native species. An exotic herbivore might however be immune to native predators.

It is tautological that if herbivores are usually predator limited, then predators are usually food limited. Except to the degree that water or some other chemical component of the food is in particularly short supply, food limitation is identical with energy limitation.

Another demonstration of relevance can be made from a combination of evolutionary theory and direct calorimetry data (Slobodkin and Richman, 1961; Slobodkin, 1961a). An assortment of whole animals was burned in a microbomb calorimeter. The observed calorific values are a skewed normal distribution with relatively low variance (Fig. 1).

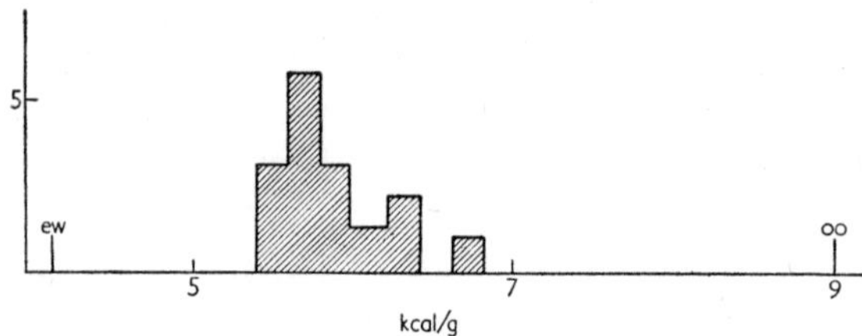

Fig. 1. The calories per ash-free gram in a collection, chosen at random, of seventeen species of animals representing five phyla. The line ew is that for carbohydrates and oo for olive oil. (From Slobodkin, 1961a.)

Table I

Variation in Calorific Value of Spitbugs (kcal/ash-free g)

Date 1960	♀	♂
27 June	5·625	5·575
14 July	6·110	6·003
2 August	6·015	5·952
16 August	6·114	5·762
5 September	5·853	5·783
13 September	5·949	5·804
27 September	5·791	5·574
Eggs 6·529 kcal/ash-free g		

Data provided by R. Wiegert (personal communication, 1961). *Philaenus leucopthalus* collected at the old field, Edwin S. George Reserve, Pinckney, Michigan.

Within a single species there may be seasonal variations (Table I). Animals about to initiate a fast are generally high in calorific value. Analysis of calorific value in pre-pupal and newly emerged adult sarcophagid flies clearly demonstrate that on emergence the flies are again at the low modal value for calorific content. The pre-pupal larvae of *Sarcophaga bullata* had 5·914, 11 day pupae 5·399 and newly emerged adults 5·079 kcal/ash-free g.

Certain species have relatively high values. The three highest values in our initial survey were those for a *Tenebrio* larva about to pupate, *Artemia* nauplii that had just hatched and were still laden with yolk, and overfed laboratory-reared *Dugesia tigrina* fed on *Artemia* nauplii.

The specimens to be burned were initially chosen at random. A tentative explanation was developed after the data from the first seventeen species (cf. Slobodkin and Richman, 1961).

We might have expected any of three possible distributions *a priori*:

1. A normal distribution, implying that energetic content of organisms is determined by the overall biochemical similarities known to exist between almost all species.

2. Broad differences between taxonomic groups might have been expected since taxonomic groups differ in so many ways that it would not be too surprising if they differed in energetic content. This is compatible with the biochemical similarities when it is considered that a biochemical potentiality does not necessarily imply the realization of that potentiality.

3. A distribution skewed in one direction is the only remaining possibility.

The observed skewed distribution with a low modal value is explained on the assumption that in general, energy is limiting to almost all populations almost all of the time and that this limitation has been the case throughout evolutionary history. This explanation is supported by evolutionary considerations since selective advantage is defined by Fisher and others in terms of the intrinsic rate of increase (m in Fisher, 1958, r in most ecological literature). That is, there is a clear selective advantage to reproduction but no clear advantage to adiposity. Excess calories will, therefore, be converted to offspring throughout evolutionary history.

If energy were not limiting, a normal distribution of calorific value in organisms or taxonomic differences in energy content might be expected. The limitation of energy would imply that animals usually maintain the lowest possible set of biochemical components consonant with survival. The possible argument that the narrow range of calorific values is due to biochemical inability to maintain other higher values is refuted by the occurrence on occasion of energy-rich organisms.

Once the above hypothesis has been formulated it would be very difficult to avoid unconscious selection of material if further testing were done by continuing our survey of animals. Calorific determinations are continuing, however, with reference to special problems.

The highest calorific value for any whole animal was 7·432 kcal/ash-free g for *Calanus hyperboreus* collected in the field by Dr R. Conover of Woods Hole. Apparently this copepod feeds very heavily on oil-rich diatoms during the brief periods of phytoplankton blooms and subsists largely on its own fat between blooms. A remarkably similar value of 7·380 was reported for a male *C. finmarchicus* by Marshall and Orr demonstrating that at least high latitude copepods become very fat and also demonstrating that the gain in calorific value is not simply an adaptation to egg laying (Marshall and Orr, 1961, personal communication).

Extremely low values were found for a razor clam, *Ensis minor* (*c.* 3·5 kcal/ash-free g), and a polychaete worm, *Strenelais articulata* (*c.* 4·7 kcal/ash-free g). These low values may be due to the inclusion of scleratized protein and polysaccharide shells, skins, and scales which would burn, thereby being included in the ash-free fraction, but would have a low calorific value. This explanation of low values requires further test.

We received samples of dried Australian brush turkey (*Leipoa ocellata*) egg yolk and found that it was identical with that of chicken egg yolk. We, therefore, burned the yolk of ten species of birds (Table II). There was no significant difference between any two species. They are all essentially identical in calorific value. Pooling all determinations, bird yolk has $8{\cdot}0 \pm 0{\cdot}1$ kcal/ash-free g. Difference between birds in precocial properties of the young and incubation period are therefore not related to yolk chemistry but to either egg size, subtle differences in developmental chemistry or both. It is also apparent that birds have long ago established their energy storage mechanism for yolk and have not been able to make it any more perfect since.

TABLE II

List of Species of Birds used for Calorimetry of Egg Yolk

Agelaius phoeniceus	*Melospiza melodia*
Archilochus colubris	*Molothrus ater*
Colinus virginianus	*Passer domesticus*
Dendroica petechia	*Phasianus colchicus*
Gallus domesticus	*Rhea americana*
Leipoa ocellata	*Riparia riparia*

Frog eggs (*Rana pipiens*), newly fertilized, are not beyond the calorific range for whole organisms (6·0) and newly-hatched tadpoles (144 hours old), with yolk still present, are in the modal region (5·8). Salamander eggs (*Ambystoma punctatum*) are identical with those of frogs. We are left with the problem of the phylogeny of energy-rich eggs in the lower vertebrates. Skate (*Raja erinacea*) egg yolk has an intermediate value of 5·6. Reptilian egg yolks are intermediate between birds and amphibians (*Urosaurus ornatus*, 6·9 kcal/ash-free g, *Sceloporus undulatus*, 6·7, *Pseudemys scripta*, 6·7, *Chelydra serpentina*, 6·6).

The calorific data abundantly demonstrate the relevance of energy analysis in ecology and actually do stimulate the formulation of evolutionary theorems and questions.

II. Theory of Energy Budgets

The relation between energetics and the numerical properties of populations must be in terms of energy budget analysis in which the population is considered as a steady-state system through which potential energy passes. We restrict attention to steady states since seasonal differences in climate and physiology combined with essentially random meteorological or biological events in short-term data collections will permit so much variance as to obscure real constancies and differences.

The concept of ecological steady states has been discussed by Odum (1957) and Slobodkin (1960) and both of these authors have indicated something of the theoretical importance of making energy measurements at or near steady-state conditions. An obvious point that neither of them mentions is the difference in meaning between the concept of steady state when applied to an entire community and when applied to a single population.

While a population may maintain its own standing crop in a steady state, no population of mortal animals can maintain a steady state in its immediate spatial environment since the process of population maintenance requires the production of a continuing stream of dead animals and, therefore, a new accumulation of potential energy in the physical environment of the populations.

To avoid confusion, let us establish units now. The population, itself, is measured in units of calories; energy income to the population or energy expenditure by the population is generally in calorie/time units. The term "cost" will be in units of calories per item, so that, for example, maintenance cost of a population will be in units of calories per calorie-days and replacement cost of an individual organism will be in units of calories per individual.

Efficiencies will always be dimensionless fractions. Only if the units of both numerator and denominator are the same can two efficiencies be legitimately compared.

There are three equations that have been commonly used to represent the energy budget of populations. They all meet the requirements of energy conservation but they differ seriously in emphasis and the translation between them might well be made explicit.

The simplest energy budget is derived by equating the energy income I, to the heat loss, R, a function of respiration, plus the yield from the population of potential energy in the form of dead animals and excretory products, Y. This has been used by many workers, including: H. T. Odum (1957), E. P. Odum and A. E. Smalley (1959), Richman (1958), and Teal (1957).

$$I = R + Y \tag{1}$$

This formulation ignores the standing crop of the population, and the composition of the yield. It is certainly adequate as a description but is relatively low in certain kinds of predictive power, since, although, yield and respiration are additive, if I should choose to remove an additional calorie per day of yield from a population, I could not reasonably expect that R would decrease by one calorie while everything else stayed constant. There would probably be changes in the size of the population, its age-structure and the availability of other kinds of yield. Up to certain limits which will be discussed below, I could actually increase yield by one calorie, with a compensatory decrease in heat production but the equation would not supply me with the technique for this.

If standing crop, P, is of primary interest, it is possible to write the following equation:

$$I = cP \tag{2}$$

in which c, the maintenance cost, is $\frac{R+Y}{P}$. Yield and respiration are obscured in this formulation but it has the advantage of permitting a solution from standing crop data and also permits some theoretical expansion that is not available to Eq. (1) alone. Maintenance cost has been evaluated using this equation for *Hydra* and *Daphnia* in the laboratory (Table III).

Either of these equations can be used to describe a community as well as a population. In communities, Eq. (1) becomes

$$I = \sum R_i + \sum Y_i \tag{1'}$$

where only the potential energy that leaves the community completely

TABLE III

	Daphnia pulex	*Hydra littoralis*
Maintenance cost (cal/4 cal-days)	1·68	0·79
Population efficiency		
adults	0·48	
young	0·04	0·06
eggs	0·06	

Values for maintenance cost and population efficiency calculated from Eq. (14) using laboratory data.

and not that which is eaten by another member of the community is counted as yield.

Equation (2) becomes

$$I = c_i P_i \tag{2'}$$

with no mention of yield. Yield production is a piece of maintenance cost from the standpoint of the species producing it and any consumption of yield from one species by another species in the community will be reflected in the maintenance cost of both species.

The expansion of Eq. (2) to the community level presents operational difficulties. Operationally, c can be evaluated from a situation in which I and P are both known for a particular species. If one is dealing with a mixed species system, it is possible to solve for the combined maintenance cost of all species in a corresponding way but the assignment of maintenance cost to each species requires at least as many different communities as there are species. There is reason to expect that any two communities with the same species composition will also have the same relative numbers of the various species unless physical conditions are different (Slobodkin, 1961b). We can also expect that change in physical conditions will alter c. We can, therefore, not evaluate the c values for each species in any particular community if the only available data are from communities with identical species lists. Solutions are possible by least squares analysis of standing crop data from communities which differ in species lists, but direct evaluation of c has not yet been made in any natural community.

If primary attention is focused on the yield from a particular population, the energy budget can be written as

$$I = \frac{Y_i}{E_i} \tag{3}$$

where the Y_i are specific kinds of yield calories and each E_i is a growth efficiency (or inverse of cost per calories of producing yield of the sort i).

E_i requires further elucidation. The energetic cost of producing an animal of age i is the total food energy consumed by that animal during its free life plus the energy expended by its parents on its behalf between the moment of its inception and its freedom. The first of these cost components is relatively simple to determine. The second is more difficult but has been done in at least one case (Armstrong, 1960) which will be discussed below. The growth efficiency up to any age is, therefore, the calories of standing crop represented by the body of an animal of that age divided by the energy expended in the animal's production (cf. Fig. 2).

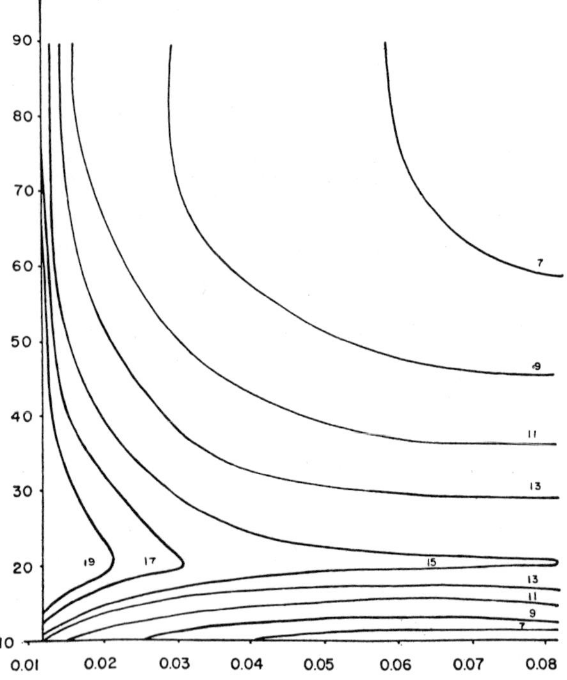

FIG. 2. Percent growth efficiency (E_i) of *Daphnia pulex* as a function of algal concentration in thousands of *Chloamydomonas* cells per ml (ordinate) and calories in the body of the *Daphnia* (abscissa). (From Slobodkin, 1960 using data of Armstrong, 1960.)

On the community level, Eq. (3) is only of significance if total yield from the community as a whole is of significance, that is, if the entire community is being treated as a device for producing potential energy.

In Eq. (3), for example, the Y_i may represent simply the dead animals produced by the population in the complete absence of predation. The total cost of this rain of dead animals is the same as the cost of replacing these dead animals with a new supply of dead.

The advantage of Eq. (3) is that, in combination with Eq. (2), it is most fruitful for the analysis of the effect of environmental change. Consider that some new source of mortality, either a predator, exploiter or disease appears. There will be a period of transition in the population and some new steady state will be achieved (except if the population is completely eliminated).

That is, assuming I constant, we will have from Eq. (2)

$$I = P'(c + \Delta c) \tag{4}$$

where Δc is the maintenance cost induced by the new mortality source. Let Y' represent the new distribution of potential energy in yield. It follows from Eqs. (2), (3) and (4) that

$$\Delta c = \frac{I}{P'} \sum \frac{Y'}{E_i} - \frac{I}{P} \sum \frac{Y_i}{E_i} \tag{5}$$

This is developed explicitly in Slobodkin (1960).

Let us consider that the change in environment involves a new predator as exploiting agent and let us assume that this exploiter is interested in proper conservation practices with reference to the population in question. He would like to know the degree to which his exploitation programme alters the standing crop of the exploited population, which is in a sense equivalent to knowing how much of the energy income to the exploited population is being diverted to the production of his yield, rather than to the production of the other sorts of potential energy-rich particles that are involved in population maintenance. He is also concerned with the energy per unit time of his yield.

He would, therefore, like an equation for the population's energy budget which will include both his yield and the standing crop, ignoring other possible yields.

The energy diverted from standing crop maintenance is $\Delta c P'$. His yield is Y_i. We define

$$E_{pi} = \frac{Y_i}{P' \Delta c} \tag{6}$$

We will call E_{pi} population efficiency. Arranging his exploitation programme so as to maximize E_{pi} will give the exploiter his most appropriate exploitation procedure since he will then be getting the maximum yield per unit depletion of standing crop. The energy budget which is of primary interest to him is

$$I = P'c + \frac{Y}{E_p} \tag{7}$$

or, if he also takes yields of the sorts $j, k, l \ldots$

$$I = P'c + \sum \frac{Y_i}{E_{pi}} \tag{8}$$

where E_{pi} (population efficiency) is given by

$$E_{pi} = \frac{P_i}{P'(c + \Delta c) - \dfrac{cP'}{P}} \tag{9}$$

Equation (8) represents a particularly interesting energy-budget equation for a single species since it combines certain properties of all three energy-budget equations. Since yield is consumed within a community, Eq. (8) reduces to Eq. (2′), for complete communities.

In summary, the various equations that have been utilized for energy studies in ecology can be intertranslated in a straightforward manner. They differ primarily in the kind of data used and in emphasis.

III. Entropy and Information in Ecology

A review of energy relations in ecology can be written with suitable incorporation of all relevant data, without ever mentioning either entropy or information in their rigorous meanings. Several recent authors have, nevertheless, felt it of value to discuss ecological energetics in terms of entropy and information. Since the theory of information has been developed, specifically, to deal with communication problems, such as determining which of a particular set of messages was actually transmitted through a communications channel which was not perfect, it is immediately adaptable to situations in which the investigator's concern is with the distribution, organization, number or arrangement of entities in an imperfectly understood situation about which he has some partial knowledge.

Margalef (1958), Hairston (1959), MacArthur (1960), and MacArthur and MacArthur (1961) have used communication theory in this way to great profit. The significance of their work has been discussed at some length by Hutchinson (1959) and Slobodkin (1961a,b).

Occasionally, information theory has been used as an analogy to suggest models that might be of ecological interest (MacArthur, 1955).

There exists a certain formal correspondence between the rigorously defined concept of information and the rigorously defined concept of entropy.

In particular

$$H = \sum_{0}^{n} P_i \log P_i \tag{10}$$

represents the information H in a set of n independent messages, each with a probability P_i of being transmitted.

An expression of the same form is used in the definition of the statistical mechanical concept of entropy.

The statistical mechanical concept of entropy is in principle equivalent to the thermodynamic concept of entropy and changes in entropy are measurable for chemical systems at known pressures and temperatures by using the relation.

$$\Delta F = \Delta E - T\,\Delta S + P\,\Delta V \qquad (11)$$

in which P is pressure, T absolute temperature, ΔV and ΔF are changes in volume and free energy, ΔF is the change in energy level of the system, defined as $Q - W$ where Q is the heat evolved during a transformation and W is the work done.

ΔS is defined as $\frac{Q}{T}$ or entropy change. The system is assumed to be thermodynamically isolated.

It would be very nice if we could, by suitable measurements, measure the various terms in Eq. (11) and, thereby, utilize the full theoretical power of thermodynamics in our analysis of ecological systems. The second law of thermodynamics, which can be verbalized as follows, "In an isolated system, the internal entropy is maximum when the system is in thermodynamic equilibrium", must be considered applicable in some sense to ecological communities. Apart from other theoretical and operational difficulties, which we will discuss below, an immediate problem arises from the fact that an ecological community cannot in any sense be considered as thermodynamically isolated, nor can any system containing a living organism be considered in thermodynamic equilibrium.

The equivalent law for a non-isolated steady state system is Prigogine's theorem which has been stated as follows by Foster *et al.* (1957): "In an open system, the rate of internal entropy production, which is always positive, is minimized when the system is in a steady state." An open system is defined by these authors as one which exchanges both energy and matter with the ambient universe. They, then, made a theoretical analysis which is immediately germane to the problem of the relation between thermodynamics and ecology.

They considered an electronic circuit in which internal entropy production is simply and directly proportional to the heat production or power dissipation by the resistance and is directly calculable from Kirchhoff's Laws. They find that for certain simple circuits the open system second law of thermodynamics actually does hold.

However, if feedback occurs within the circuit, Prigogine's theorem does not necessarily hold. If the system is characterized by the presence of interlocking feedback loops, the theorem only holds if arbitrary restrictions are introduced. Even for the case of uncrossed or non-interlocking feedback loops, the theorem is only valid if the power source for the feedback loop is contained within the system.

The degree to which an ecological community can be analogized to an electronic circuit is arguable (see Slobodkin, 1960), but it is clear that ecological communities are feedback systems of high complexity in which the power source for the feedback components, even if they could be physically distinguished in the way an electronic feedback component can, is almost certainly external to the system. For ecological communities, it is, therefore, impossible to make any unequivocal statement at all about the relation between steady state conditions and the rate of entropy production. Obviously net cosmic entropy is increased by the activity of ecological communities, but this is not a particularly surprising or heuristic conclusion.

Foster, *et al.*, continue with a general analysis of the limits of applicability of Prigogine's theorem but this is not of immediate ecological concern except to note that they were unable to find any other thermodynamic property that could be theoretically demonstrated to reach either a maximum or minimum when any complex feedback system comes to a steady state.

It might be noted concurrently that if the mass of an open system stayed constant, and if the rate of entropy production came to a minimum, the total entropy of the open system must also come to a minimum. While ecological communities may meet the first condition, we have no reason to believe they meet the second. The often repeated statement that evolution tends to lower the entropy of living organisms is not clearly demonstrated and is of problematic value.

Therefore, the most interesting theorems of thermodynamics don't seem to apply to ecological systems in any direct way. The interest of translating directly measurable ecological parameters into the language of thermodynamics is not obvious. Nevertheless, several authors have attempted this translation and have produced conclusions which might at first glance be mistaken for empirical generalizations. The two major recent expositions of the application of thermodynamic theory to ecology are those of Patten (1959), and Odum and Pinkerton (1955).

Patten quotes the aphorism of Schrodinger (1946): "What an organism feeds upon is negative entropy. Or, to put it less paradoxically, the essential thing in metabolism is that the organism succeeds in freeing itself from all the entropy it cannot help producing while alive."

Schrodinger points out, in support of the notion that negative entropy is what is consumed by organisms, the apparent lack of logic in metabolism. That is, "Any atom of nitrogen, oxygen, sulphur, etc., is as good as any other of its kind. What would be gained by exchanging them?" and also, "For an adult organism, the energy content is as stationary as the material content since, surely, any calorie is worth as much as any other calorie. One cannot see how a mere exchange could help" (Schrodinger, 1946).

This is not merely a jocular or trivial point. It is possible to conceive a world in which organism-like entities do actually require only enough food to make good entropic gain and do not replace existing biomass. It is equally possible to imagine an astronomical world in which planets follow the eminently logical paths of epicycles. The reason for an organism requiring energy may be obvious, but after Schrodinger's question, it may deserve restatement. Organisms are not exchanging one calorie for another nor are they only maintaining body heat and performing the other energy-utilizing operations of normal physiology. When the first animal ate its first plant, the animal was not exchanging, it was gaining. To meet the conditions set by natural selection, the plant had to increase its rate of incorporation of energy to make good the loss to the animal or it would have disappeared in the process of evolution. Animals need energy to make good their loss of energy to other animals. We're not simply dealing with a steady state system sucking in negative entropy to maintain itself against the laws of thermodynamics, but we have a whole set of such systems, each one acquiring energy and matter at the expense of other organisms to make good its losses to yet other systems. The only reason for this state of affairs is that the systems or organisms which behaved in a rational way, as if they understood Schrodinger, have long since been eliminated by natural selection.

The energy losses of an organism in a population are not simply heat. Corpses, faeces, exudates are all necessary by-products of evolutionary success. The rate of energy passage through an organism or population can, in fact, be altered by altering the predation rate, and this is different from increasing the population's heat production. If a closed system including a living organism is considered, the metabolic activities of the organism in maintaining itself in an unchanged condition result in an entropy increase in the closed system. If it were possible in principle to measure the entropy of the isolated organism itself, this would have been found to be unchanged. Therefore, the organism is acting to increase the entropy of the world around it.

Patten carries the paradoxical part of Schrodinger's statement further by stating that living organisms feed upon negative entropy to compensate for information losses attending the life process (Patten, 1959). Here the formal similarity between information and the statistical mechanical concept of entropy is taken to demonstrate identity between information and negative entropy. However, information in communication theory does not have temperature as a significant parameter nor can the complete array of states be specified so as to permit a statistical mechanical definition of entropy to be operationally evaluated in any biological system. In short, while certain analogies between parts of ecological energetics and parts of thermodynamics can be verbalized, there is no evidence whatsoever that these are necessary or even fruitful for the advance of ecological comprehension.

Not only are thermodynamic analogies current in the literature but circuit diagrams and hydrostatic flow diagrams are also taken as analogies. All of these violate common sense. Note the statement by H. T. Odum (1960) in discussing an analogy between electric circuits and ecological communities: "The validity of this application may be recognized when one breaks away from the habit of thinking that a fish or bear, etc. takes food and thinks instead that accumulated food by its concentration practically forces food through the consumers." To my knowledge this sort of analogy has produced neither suggestions for practical experiments nor significant syntheses.

IV. Lindeman's Theoretical Formulation

The classical initial study of energy-passage through a natural community is that of Lindeman (1942).

The framework into which Lindeman fitted his data was essentially the following. Assume all organisms in a natural community to belong to one and only one of the trophic levels designated by $\Lambda_1, \Lambda_2, \Lambda_3 \ldots \Lambda_n$ such that any organism at trophic level $\Lambda_{i>1}$ is nourished by eating organisms of trophic level Λ_{i-1}. Trophic level Λ_1 consists of autotrophs deriving their energy from the sun. The energy passed per unit time from trophic level Λ_1 to trophic level Λ_{i+1} is designated as λ_i and is referred to as the productivity of level Λ_{i+1}. A problem has risen in the literature about whether the food consumption or the protoplasm synthesis of Λ_{i+1} should be called its production but this problem is not of fundamental importance for our immediate purpose. I'll try to keep the concepts clear as we proceed. Λ_i is a standing crop with a dimension of calories, λ_i has a dimension of calories per time and both are calculated per cm^2 of surface.

There is a relation between the concept of trophic levels and that of food chains but the two concepts are not identical, the first being a simplifying assumption while the second is purely descriptive. If a diagram of the passage of all high-energy molecules through an ecological community is made, it will be found that the potential energy of any given molecule will either have dissipated as heat in or near the body of some organism or been transferred to some other organism. The individual organisms can be arranged by drawing arrows to a point representing any organism from the points representing all other organisms from which it has at any time received a high-energy molecule. The resultant network of arrows is a food chain. Since objective taxonomic criteria of specific discreteness exist, it is possible to superimpose all points representing organisms of the same species, which considerably simplifies the diagram. The concept of trophic level is based on the assumption that in any food-chain (or food-web) diagram there exist classes of points, each class being defined by a constant number of arrows intervening between any point in the class and some initial point characterized by the absence of arrows directed towards it (i.e. an autotroph). This also carries the implication that there is always a fixed number of arrows, or food-chain links, in the passage from an autotroph to any particular species, regardless of the route chosen. If this assumption is met, then all points characterized by a constant number of arrows (i) between them and an autotroph can be collapsed into a single point. The mean standing crop of all species represented by this single point is called Λ_{i+1} since autotrophs are Λ_1. All arrows leading to this point can be collapsed into a single arrow and the total energy flow represented by this arrow (in cal/time) is referred to as λ_i.

Several questions are raised by this formulation and the attempt to answer these questions empirically and theoretically has occupied most of the workers in the field of ecological energetics ever since Lindeman's paper appeared.

1. Is there any maximum number of possible links in a food chain? In a stronger form, we could ask, is there any characteristic number of links in a food chain.

2. Is there any characteristic ratio between standing crops of species at different locations in a food chain? In the terminology of trophic levels does knowledge of the indices i and j of two trophic levels predict in any sense the ratio $\Lambda_i : \Lambda_j$.

3. Are there any constancies in the ratios of the productivities of a species and the predators feeding on it? That is, is $\frac{\Lambda_j}{\Lambda_i}$ a constant? This ratio has been called either food-chain efficiency or ecological efficiency by other authors (Slobodkin, 1959, 1960; Englemann, 1961).

4. Are all or any of these questions interdependent? For example, could the existence of a characteristic number of links in a food chain permit prediction of the answer to the other questions.

These questions are independent of the simplifying assumptions made by Lindeman and also independent of the criticism of Lindeman's work which will be stated below.

Since food-chain efficiency is clearly less than one, the greater the number of links in the food chain the lower the energy income per cm^2 of earth surface per time of the organisms high in the chain. If food-chain efficiency is constant (say E), the energy income of any species will be proportional to E^i where i is the mean number of food-chain links between that species and the autotrophs.

A species high in the food chain (j) might have an abundance equal to that of a species low in the food chain (i) if the ratio $\frac{c_j}{c_i}$ is equal to $\frac{c_i}{c_j}$ or greater than $E^{(j-i)}$ where E is the constant food-chain efficiency and c is maintenance cost. Since a large part of the maintenance cost of any species is respiration, and since there is no reason to expect a hunter to do less work than its prey, we would not expect species high in the food chain to be as abundant per unit area as those low in the food chain. Should maintenance cost of species be constant, we would expect the abundance of species at levels j and i to be proportional to $E^{(j-i)}$. The classical Eltonian pyramid depends on the fact that typically there is a correlation between body size and trophic level. Reversal or inverted Eltonian pyramids occasionally occur either as a temporary distortion of the normal steady state (Evans and Lanham, 1960) or as a consequence of extremely heavy predation and rapid growth in the lower levels of the food chain (Odum and Odum, 1955).

The maximum number of possible links in the food chain is dependent on relative abundance as a function of food-chain position. Since λ is expressed as cal/area, the food for an animal sufficiently high in the food chain is so dilute as to place it in the position of the sheep which must run, not walk, between grass blades lest it starve to death. The questions raised by the Lindeman formulation are, therefore, intimately interrelated.

There is, however, a serious logical error in Lindeman's study which effectively invalidates all his estimates of productivity and efficiency. This is not simply a matter of slight differences in definition or of high variance in the initial estimates. Since his procedure has been followed by other authors (Dineen, 1953) and his estimates have been quoted in various contexts (Slobodkin, 1960; Patten, 1959; and others) it is of importance to prevent further reliance on these data.

As already indicated, the energy budget for a trophic level can be written as

$$I = \text{Respiration} + \text{Yield}$$

Yield consists of all potential energy leaving the trophic level, including that consumed by predators and decomposers.

Lindeman, however constructed the following energy budget

$$I = \text{Respiration} + \text{Yield} + (\text{Turnover time} \times \text{Standing crop})$$

The final term is superfluous, regardless of the definition of turnover time.

It is not safe to salvage any of the values from Lindeman's data due to his sequential procedure of evaluation. The same remarks apply to the data of Dineen (1953) and to Lindeman's analysis of the data of Juday (1940).

It is of interest that Clarke (1946) does not commit the error of Lindeman but does not point it out explicitly.

The questions raised by Lindeman remain valid and to a large degree unanswered.

V. Daphnia Energetics

Perhaps the most complete study of ecological energetics has been made with *Daphnia* populations in the laboratory. Pratt (1943) demonstrated that *Daphnia magna* populations in the laboratory will fluctuate in even a constant environment. Slobodkin (1954) confirmed this result of Pratt's and showed that size of *Daphnia obtusa* populations in the laboratory is linearly dependent on food supply. The population fluctuations were considered to arise because of age and size-specific differences between the individual *Daphnia* composing the population. A population composed primarily of small, young animals will have a lower feeding rate than a population of the same number of large animals. Growth and reproduction in *Daphnia* are closely dependent on food supply. Under starvation conditions reproduction and growth effectively cease. Mortality is not severely altered by changes in nutrition (Frank, 1960). Given appropriate age and size distribution, mortality reduces competition for food just sufficiently to enhance the reproductive and growth rates of the survivors and this permits the population to return to its initial age structure. It can be shown in theory (Slobodkin, 1961b) that for a species with an essentially rectangular survivorship curve the number of animals of a given age in this stable age-structure is proportional to the inverse of the growth rate at that age. Age-structure change combined with the maintenance of severe starvation and the fact that different sized animals have different food consumption rates requires numerical fluctuations in the population even if the environment is kept as constant as possible.

If animals are removed from *Daphnia* populations by the experimenter at some fixed rate, the size of the residual population is reduced while the linear dependence on food supply persists. When small animals are preferentially removed (see Fig. 3) the relation between population size P_f and removal rate F (expressed as number of animals removed per unit time divided by the births during that time) is given by the simple equation

$$P_F = P_0\left(1 - \frac{F}{2-F}\right). \tag{12}$$

This relation holds only to a first approximation when adult *Daphnia pulex* are removed (Slobodkin, 1959, and Fig. 4) and is also approximately valid when the experiment is repeated with *Hydra oligactis* populations (Slobodkin, 1961a) but attempts to reproduce these results

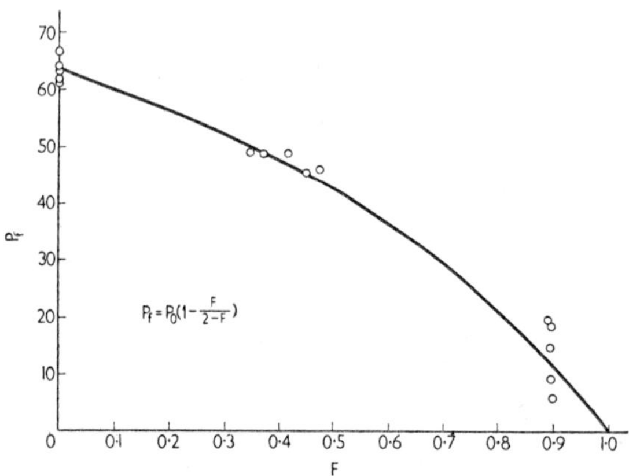

Fig. 3. The relation between size of residual *Daphnia pulex* population (P_f) and removal rate as fraction of newborn (F) when young animals are preferentially removed. The line is drawn using Eq. (12).

with a stable two species system consisting of *Hydra littoralis* and *Chlorohydra viridissima* have not been successful. Armstrong (1960) has found that Eq. (12) does not hold in *Daphnia pulex* for negative values of F (i.e. immigration). The equation can, therefore, be considered as representing one of the simplest of a series of possible equations of the general form.

$$P_f = P_0\,(1 - \phi\,(F)\,) \tag{12'}$$

where $\phi\,(F)$ involves the compensatory mechanism of the population in its response to predation and its own density, either as an explicit biological model, in which all relevant parameters are separately evaluated and appropriately combined (as recommended by Watt, 1960) or simply an empirical approximation.

Richman (1958) made calorific determinations of *Daphnia pulex* of different ages and reproductive conditions. He also determined the calorific value of *Chlamydomonas reinhardi* which was used as food by the *Daphnia*. First and second instar animals and pre-adult animals had 4·1 kcal/g while parthenogenetic females were considerably fatter (5·1 kcal/g). *Chlamydomonas reinhardi* produced 5·3 kcal/g (5·5 kcal/ash-

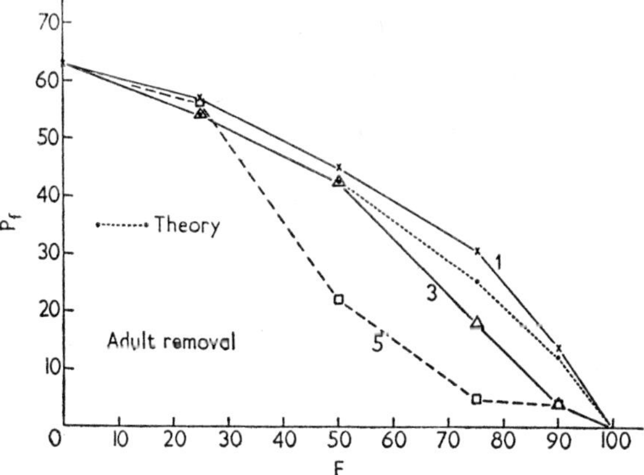

FIG. 4. The relation between P_f and F when adult *Daphnia* are preferentially removed. The numerals 1, 3 and 5 refer to food levels. There is deviation from Eq. (12) which is more marked at high food levels and high values of F.

free g). Ketchum and Redfield (1949) analysed six species of Chlorophycae and found a total calorific range of 5·1–5·5 kcal/g (5·9–6·4 kcal/ash-free g), indicating that neither the *Chlamydomonas* nor the *Daphnia* are unusual in this respect.

Richman then demonstrated that filtering rate in *Daphnia* is independent of algal concentration over a fourfold concentration range and evaluated the filtration rate of different size *Daphnia*. The filtration rate per animal increases with age but algae acquired per mg of tissue falls off. A young animal is in this sense better nourished than an old one, which may explain to some degree the higher growth rate in small animals. From the above data, Richman could compute the growth curve and energy consumption of *Daphnia*.

He then determined oxygen consumption ($6 \cdot 5 \pm 0 \cdot 7$ μl/mg/h dry-weight) for various sizes and conditions of *Daphnia*. These values agreed in general with other values in the literature for planktonic Crustacea ($7 \cdot 2 \pm 4 \cdot 3$ μl/mg dry-weight/h O_2 for seventeen species of planktonic Crustacea). R.Q. for *Daphnia* was also measured and found to vary from *c.* 0·75 for starved animals to *c.* 1·00 for fed animals. Energy expenditure could be computed by taking the consumption of 1 ml oxygen as equivalent to 0·005 cal heat produced (the calorific equivalent of oxygen consumption varies only 5% over the R.Q. range 0·71–1·00).

Richman defined the energy of growth as the calorific equivalent of the protoplasm added to the animal and defined the energy of reproduction as the calorific value of the young animals produced. These plus the energy equivalent of respiration were assumed equal to the total assimilated energy during the time interval of measurement. The energy ingested was determined from the calorific value of the *Chlamydomonas* and the observed feeding rate. Non-assimilated energy was determined by difference. These determinations were made separately on pre-adult *Daphnia* and reproductive *Daphnia*. In the pre-adult animals, growth can be seen separate from reproduction.

Armstrong's (1960) growth-efficiency estimates involve extrapolating Richman's data for the pre-reproductive period into adult life. Richman's growth-efficiency is calculated either as $\frac{\text{new protoplasm}}{\text{energy consumed}}$ per unit time which is the gross growth-efficiency or energy coefficient of growth of the first order of Ivlev (1945) and Ricker (1946) or as $\frac{\text{new protoplasm}}{\text{energy consumed} - \text{energy egested}}$ per unit time which is the net growth-efficiency or energy coefficient of growth of the second order as defined by these authors.

Gross growth-efficiency varies with algal concentration from 13% at low algal concentration ($2{\cdot}5 \pm 10^4$ cells/cm^3) to 4% at high (10 ± 10^4 cells/cm^3) but net growth-efficiency was constant at from 55% to 59%. The speed of digestion was apparently limiting in this system. It seems likely that with sufficient dilution of the food, efficiency would decrease, since the effort expended in filtration is constant while the energy return from the filtration process is proportional to food concentration (Armstrong, 1960; Slobodkin, 1954).

Growth-efficiency was low after attainment of reproductive age (less than 1½% gross or 4% net). Gross reproductive-efficiency was of the same order as pre-adult growth-efficiency (*c.* 10–17%) and was similarly dependent on algal concentration. Net reproductive efficiency, however, increased with feeding rate from 52% to 70%. The reasons for this increase do not seem obvious unless we consider that a certain constant amount of ingested food is relegated to maintenance of the animal while the remainder is utilized for growth and reproduction. This conforms with the generally accepted concepts of fish growth (Beverton and Holt, 1957).

Richman presented the following energy budgets at the four algal concentrations summarizing forty days of individual growth.

cells × 10^3/ml	I =	(growth + reproduction) +	respiration +	egestion
25	6 140	1 070	841	4 229
50	13 586	1 774	941	10 871
75	20 739	2 354	1 020	17 365
100	29 238	2 928	1 084	25 226

Armstrong, using eight experimental populations and the calorific values of Richman, computed growth-efficiency and reproductive-efficiency by a least squares fit to the energy budget:

$$\text{calories ingested} = \frac{\text{calories of growth}}{\text{gross efficiency of growth}} + \frac{\text{calories reproduction}}{\text{gross efficiency of reproduction}}, \quad (13)$$

finding growth-efficiency as 7% and reproductive-efficiency as 5%. Armstrong estimated the mean cell concentration in his populations as approximately $10^4/\text{cm}^3$. As indicated above, efficiency may be expected to decrease at sufficiently low algal concentrations.

Building on the work of Richman, Armstrong (1960) has made a particularly interesting analysis of growth-efficiency and age in *Daphnia* in which he considers the origin of age to be at the onset of production of the egg which will give rise to the animal in question.

Armstrong assumed that the growth-efficiency determined for the period immediately prior to the onset of reproductive activity persisted into reproductive life but that the total amount of energy available for growth had been reduced in favour of reproduction. The total energy associated with growth during reproductive life is, therefore, the total calorific content of the growth-increment divided by the growth-efficiency at the termination of pre-reproductive life.

The cost of producing a single young animal is calculated by Armstrong as the calorific content of newborn divided by reproductive efficiency. This cost is added to the food consumption during free life to form the denominator of Armstrong's growth-efficiency calculation. The calculated relation between growth-efficiency, animal size and algal cell concentration as calculated by Armstrong is summarized in Fig. 2.

Using the calorific determinations of Richman, Slobodkin (1959) could state with reasonable accuracy for twenty-eight *Daphnia* populations the calorific equivalent of the food, standing crop and, for twenty-three of these populations, the yield in calories of young, adults, and eggs. Slobodkin assumed that all of his populations (P_F) were related to the size of a control population (P_0) by Eq. (12). It had previously been demonstrated by Slobodkin (1954) that *Daphnia* population-size is linearly related to food supply when food consumption equals food supply. Further, all food supplied is consumed by a *Daphnia* population in the absence of predation.

At high values of F, food was left unused in Slobodkin's populations. The amount of food consumed by each population was, therefore, estimated for each population P_f by substituting F and P_f in Eq. (12), solving for P_0', the theoretical size of a population with food consumption identical to P_F but with $F=0$. The food consumed by P_F was then assumed to be $\dfrac{P_0' \times \text{food supplied}}{P_0}$ where P_0 was the size of a population free from predation, with the same food supply as P_F.

Clearly, population-size and yield are proportional to food consumed, all other things being equal. In the absence of predation, the relation between population-size and energy consumed is given by Eq. (2), the constant of proportionality being the maintenance cost in units of calories per calorie-day. In *Daphnia* populations a mean value for maintenance cost derived from five populations is $c = 1{\cdot}69$ cal/4 cal-days. Omitting these five populations from further computations, an equation of the form of Eq. (8) was written for each of the twenty-two populations. These were reduced to four linear equations of the form

$$\sum_0^{22} Ix_i = \sum_0^{22} x_i(cP') + \sum_{j=1}^{3}\left(\sum_0^{22} x_i\left(\frac{Y_j}{E_{pj}}\right)\right) \tag{14}$$

in which x_i takes the values c, $1/E_{pj}$ successively in the four equations and E_{pj} represents the population efficiency of producing eggs, adults or young. Population efficiency is defined by Eq. (9).

Best approximations of c and E_{pj} are determined by least squares from the four equations and are presented in Table III. Note that the maintenance cost, c, determined by this procedure is essentially identical with the corresponding value determined directly from control populations ($c = 1{\cdot}68$).

As seen from Eq. (9) E_{pj} is clearly distinct in definition from the concepts of efficiency used by Richman, Armstrong or Lindeman. To any mixed predation procedure, there corresponds what might be called an overall population efficiency (E_p) calculated as

$$\frac{1}{E_p} = \sum \frac{Y_j/\sum Y_j}{E_{pj}} \tag{15}$$

or if Eq. (12) is satisfied

$$E_p = \frac{(2Y/F) - Y}{I} \tag{15'}$$

where Y is total yield per unit time in calories (Slobodkin, 1960). In an optimal predator-prey system the maximum possible value is approached. From the analysis of the *Daphnia* data and equation it would appear that removal of the largest animals, with the slowest growth rate and the highest probability of dying in the near future in any case would be optimal predation procedure. If so, and if prudence of this sort has any evolutionary significance, natural predators should display prudence in this sense. Some obviously do, by consuming only moribund, diseased and superannuated prey. It may reasonably be argued in this case that these are the only prey that can be easily caught. Ivlev indicates (1961), however, that when fish that show size-discrimination in their feeding are presented with prey of various sizes, they choose the largest individuals. It might again be argued that this is adaptation to acquiring the greatest food energy per unit energy expenditure in limiting and this could not be denied. Critical field evidence is lacking.

Examined from the standpoint of the prey, increasing the relative population efficiency of those classes of individuals that are normally taken by predators obviously is of selective advantage to the prey. This can be done most feasibly by altering the age distribution of reproductive activity so as to increase the reproductive value (as defined by Fisher, 1958) of those animals not taken by predators and minimize the reproductive value of those normally taken. The persistence in the population of old animals of low reproductive value will itself presumably be selected against to save the cost of their maintenance should predation fail to remove them. This argument may be of significance in connection with Medawar's suggestions (cf. Comfort, 1961) on the selective significance of age at physiological death. I would not be surprised if a rigorous theoretical translation between reproductive value and the inverse of population efficiency could be stated but I cannot do this at present.

There is no clear reason why population efficiency should be expected to have any particular constancy from one species or situation to the next except for a general increase with age and even this may not be monotonic.

Population efficiency was also evaluated on a set of sixteen *Hydra oligactis* populations fed on nauplii of *Artemia* sp. While the relation between F and P_f in *Hydra* did not neatly conform to Eq. (12) it is equally impossible to claim that the general applicability of the equation is denied by the *Hydra* (Fig. 5).

The population efficiency and maintenance cost of the *Hydra* populations is also indicated in Table III. Energy income for the *Hydra* was assumed to be the energy content of the *Artemia* nauplii that were provided as food. The similarity between the population efficiency of *Hydra* yield and *Daphnia* egg yield is probably fortuitous, despite the tempting hypothesis that the relatively simple anatomy of a *Hydra* is more similar to an egg of a Metazoan than to anything else. There is a great paucity of data on the metabolic cost of complexity. Unpublished data collected by Miss Jill Claridge in my laboratory indicate that the total energy expenditure of a frog embryo between the time of fertilization and the time of initial heartbeat (*c.*145 h) is less than 5% of the energy present in the newly-fertilized egg. Energy loss is considerably more rapid after the heart starts beating. If this result

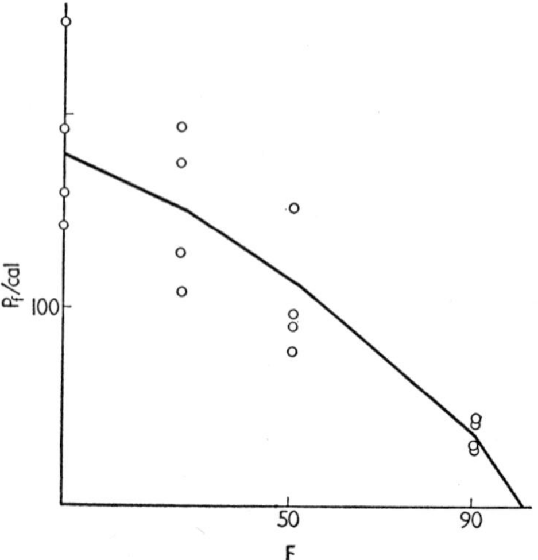

FIG. 5. The relation between P_f and F in *Hydra oligactis* populations. The line is from Eq. (12). Four multiples of a base food supply were used.

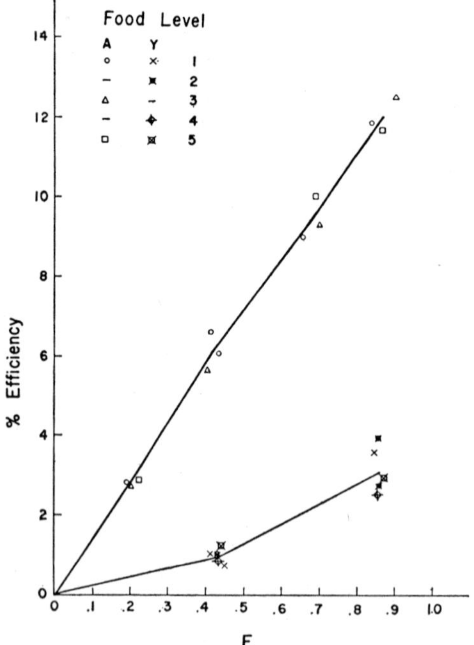

FIG. 6. Maximum estimates of ecological efficiency of *Daphnia pulex* populations as a function of F. The lower line represents removal of young.

is confirmed, it would tend to indicate that morphological complexity *per se* does not require large energy-expenditures but that the primary energy cost in organisms is that associated with mechanical work.

Ecological efficiency as the quotient (yield calories to predator)/(food calories to prey) can be readily evaluated for the *Daphnia* and *Hydra* populations as a function of predation intensity, F. The food consumption for the *Daphnia* can either be taken as the assumed value derived from Eq. (12), as in the population efficiency analysis, or can be considered equal to the food provided, free of assumptions (see Fig. 6). In the case of the *Hydra* populations, food provided was taken as the denominator. The data are presented in Fig. 7. The maximum observed values for ecological efficiency in these experiments, therefore, varies from *c.* 7% for the *Hydra* to *c.* 13% for the *Daphnia*.

These experiments are, to my knowledge, the only evaluations of ecological efficiency under relatively controlled laboratory conditions. As far as they go they suggest that ecological efficiency may either be constant or may only take a narrow range of values in nature since the taxonomic and ecological difference between *Daphnia* and *Hydra* may be as significant as the differences between any two field situations.

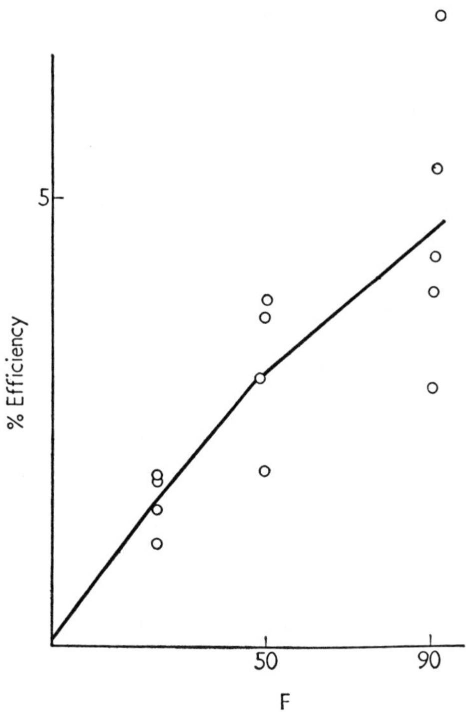

FIG. 7. Ecological efficiency of *Hydra oligactis* as a function of F

VI. Field Estimates of Efficiency

There are very few field studies which have any possibility of shedding relevant light on the value of ecological efficiency.

The steady state condition can be readily approximated in the laboratory. In the field the existence of a steady state, in the limited sense of identity of standing crop at the beginning and end of the study may be empirically demonstrable but often the deviation from a steady state is evaluated and calculations made accordingly. Various correction factors and sampling errors are involved in field data. Unless there is some apparent logical or biological impossibility involved, as in Lindeman's work, it seems best to accept field estimates at their face value in the hope that there is no consistent trend to the errors introduced.

Golley (1960) studied the energy budget of a selected group of species on an old field in Michigan, in particular *Microtus pennsylvanicus* and *Mustela brevicauda*.

The magnitude of the *Microtus* values hinges on an assumed factor which translates the number of animals trapped into population density. Given this qualification, the energy budget was of the form of Eq. (1) in which I was taken as the sum of growth and respiration, that is faeces were assumed to be composed entirely of unassimilated material.

Golley's best estimate of respiratory loss as (respiration energy/energy consumed) is 68% for *Microtus* and 93% for *Mustela*. The estimates are based on fasting metabolism and may underestimate respiratory losses under conditions of normal activity.

Faecal material constituted 10–18% of the energy consumed by *Microtus* and 10% by *Mustela*.

The *Mustela* energy budget is complete in that predation on *Mustela* is essentially negligible and faecal and respiratory loss seem to balance energy consumption. For *Microtus*, however, 14–22% of the energy consumed remains to be accounted for. The energy loss associated with body activity above the resting level must be subtracted from this although the precise percentage cannot be determined.

Microtus is considered the primary food supply of *Mustela*. The ratio of food consumed by *Mustela* to food consumed by *Microtus*, is 2%. Presumably, other predators also feed on *Microtus*, which would somewhat increase the estimate of *Microtus* ecological efficiency. The range of likely values for *Microtus pennsylvanicus* in an old field is, therefore, 2–22%.

Teal (1957) studied a small leafy pond. The chief plant-nutrient source was apple leaves dropped in during autumn. Over forty species of animals were found in the spring but only 7 of these were of sufficient quantity to merit separate energy analysis.

The calorific determinations were made by a potassium permanganate reduction technique used originally by Ivlev, on a wet-weight basis.

Since most calorific determinations (Golley, 1961; Slobodkin and Richman, 1961) have been per dry-weight, these determinations are of interest in themselves. (See Table IV.)

TABLE IV

	cal/Fresh ϕ	% max. est. ecol. efficiency
Diptera		
Calopsectra dives	690	20
Anatopynia dyari larva	880	13
Anatopynia dyari adults	1 580	26
Oligochaeta		
Limnodrilus hoffmeisteri	760	—
Platyhelminthes		
Phagocata gracilis	1 330	—
Amphipod		
Crangonyx gracilis	810	—
Isopod		
Asellus militaris	—	20

Respiration measurements were made in closed containers *in situ* in the spring. Laboratory growth-studies were made for *Phagocata* and for *Anatopynia* larvae.

Mortality rates were calculated, from the difference between the increase possible for each species in the absence of mortality and the observed change in the pond, using an equation from Ricker (1946)

$$P_t = P_0{}^{e(k-i)t}$$

where k is growth rate, i is mortality rate, e is base of natural logarithms, P_0 is population weight at time zero, and P_t is population weight at time t, which in this study was one month.

The mortality rate was weighted for field population size, as determined by field samples, to determine total mortality.

The energy budget is given as:

(Respiration + all exuvia + mortality + increase in standing crop = assimilated food.)

For each species (assimilation − respiration) provides a maximum estimate for yield.

The efficiency estimates are listed in Table IV. The flatworms raise a particularly interesting problem since as top trophic level carnivores, their ecological efficiency as we have been using the term is zero by definition. Nevertheless, they produce a significant amount of slime. This slime is in one sense part of the flatworm population since it serves to entangle food organisms and to facilitate the locomotion of the worms. It almost certainly serves as food for various bacteria, and, considering the catholic taste of amphipods and isopods, may serve to feed them also. We might, therefore, consider the slime as yield from the flatworm. If we did, we would also, of necessity, have to consider non-predatory mortality as constituting yield for the same reason. With mortality estimated at 48 kcal/m^2/year, mucus production at 89 and energy income at 131, we would have a fantastically high efficiency for flatworms.

This forces attention on a logical problem associated with our original definitions of yield and efficiency. In the *Daphnia* experiments, if the experimenter did not take any yield, the ecological efficiency was considered to be zero. There was, nevertheless, potential energy leaving the populations as non-predatory mortality, faecal material and cast skins. Given an appropriate bacterial flora, we might then have considered the energy relations of the system algae — *Daphnia* — bacteria. Slobodkin (1959) calculated that 5% of the consumed algae is accounted for by mortality in the absence of predation in *Daphnia*. Predation of any kind competes with other sources of mortality and at maximum predation intensity it seems almost possible to completely eliminate non-predatory mortality.

The logical problem is whether or not to include energy-rich exudates, wastes and deposits as yield. Arbitrarily our discussion has been confined to actively acquired yield, that is, yield which would not have been acquired by the predator without some activity on the predator's part, specifically not including activities of scavengers, bacteria and others which might consume potential energy after it has passively departed from the prey population. This has been a matter of convenience. Quantitatively, inclusion of passively-lost potential energy as yield from each population will raise the ecological efficiency estimates to varying degrees. The maximum increase in the case of *Daphnia* will be somewhat greater than 5% at low active predation levels and somewhat less at high predation levels. The 5% estimate of Slobodkin (1959) did not include cast skins. High predation lowers the rate of non-predatory mortality. Determination of yield by Teal's technique of substracting respiration and standing crop change from energy income provides a maximum estimate which will include both actively acquired yield to predators and the yield to scavengers and bacteria.

Steady state conditions were not demonstrated by Teal but the relative permanence of the spring contours, taken in conjunction with the leaf fall would indicate that the values measured are not merely transient stages of a rapid successional process.

Odum and Smalley (1959) have presented a brief summary of Smalley's doctoral disseration on the energetics of a grasshopper (*Orchelimum fidicinium*) and a snail (*Littorina irrorata*) on a Georgia *Spartina* salt marsh. Assimilation was estimated from the sum of the calorific equivalents of respiration and total growth which had occurred in the populations during the study period. Respiration was determined in the laboratory.

The role of predators in mortality was not evaluated for either the snail or grasshopper. Year class phenomena seem to occur in the snail; indicating a strong possibility that the steady state condition was not met. Assuming the estimated total growth increment to be yield, in some sense and taking estimated food energy assimilated as the denominator, the ecological efficiency of the snail is 14% and of the grasshopper 37%. If food ingested is taken as denominator the efficiencies are 6% for the snail and 13% for the grasshopper.

Odum and Smalley note that grasshoppers feed on *Spartina* and grow rapidly, while the snails feed on detritus and grow slowly. The total energy assimilated by the snails is only about twice that assimilated by the grasshoppers although the summer average standing crops were 700 snails/m^2 and 10–20 grasshoppers/m^2.

H. T. Odum (1957) in a long paper analysing the energetics of a warm constant-temperature spring (Silver Springs, Florida) has found the order of magnitude of ecological efficiency to be from 5 to 16%. Odum's detailed theoretical analysis is too complex for adequate discussion here, but his conclusions are in general agreement with those of other workers.

Englemann (1961) has combined laboratory and field studies in an analysis of arthropod microfauna energetics in a Michigan oil field, with emphasis on the Oribatid mites.

Feeding experiments were conducted in the laboratory, using fungi and yeast as food. Respiratory rates and growth rates were measured directly. Bomb calorimetry was used to determine energy equivalents of the mites' food.

He constructed the following energy budget for the oribatid mites in a square metre of old field.

Ingested	Faeces	Egg mortality	Adult mortality	Respiration
10·25 kcal	7·69	0·16	0·27	1·97

All of these estimates were independent of each other in the sense that none were determined by difference. It is, therefore, very remarkable that the total discrepancy between the input estimate and the estimate of expenditures is only 0·16 kcal/m^2/year. Ecological efficiency (mortality/ingestion) is 4·2% if all mortality is assumed due to predators. I believe this to be as trustworthy an estimate as we have for a field population. Dividing standing crop into ingestion provides a maximal estimate of maintenance cost of 38 cal/cal-year, or assuming additivity of maintenance cost with time, 0·42 cal/4 cal-days which is lower than that for *Daphnia* or *Hydra* but of the same order of magnitude.

Englemann also attempted to compute ecological efficiences for whole trophic levels (defining trophic level in the sense of our previous discussion). The energy sources of the soil community are the corpses and faeces deposited on the soil surface by the above-ground organisms. The plant base of the subsoil community is the fungi. These are eaten by herbivorous mites and collembolans and these in turn are consumed by carnivorous mites. Other animal groups (for example nematodes) occur but are not of quantitatively major significance. Englemann pooled the available data for respiration, standing crop and food habits of Collembola, Protozoa, Symphyla, Pauropoda, Japygidae and the Oribatidae and followed the procedure of defining ecological efficiency of the herbivores as

$$\frac{\text{cals ingested by carnivores}}{\text{cals ingested by herbivores}}.$$

Not knowing ingestion directly he derived ecological efficiency by taking

$$\frac{\text{Respiration cals of carnivores}/\%\text{ assimilation by carnivores}}{\dfrac{\text{Respiration cals of herbivores}}{\%\text{ assimilation by herbivores}}+\dfrac{\text{Respiration cals of carnivores}}{\%\text{ assimilation by carnivores}}}$$

(Note that this was incorrectly printed in his paper but was used correctly in his calculations.)

The implicit assumption here is that there are no secondary carnivores. Assimilation rates were not directly measurable nor was the diet of each species known with certainty. Englemann, therefore, calculated ecological efficiency using all possible self-consistent combinations of assumed diets and assimilation rates. He concluded that any value from 8% to 30% would be an acceptable estimate of ecological efficiency given present data.

VII. Conclusions

It is clear from field data, evolutionary theory, direct calorimetry data and a simple theoretical analysis that energy is of major significance in ecological systems.

It is possible to show that the various theories that have been utilized by workers in the field of ecological energetics are mutually intertranslatable.

The theoretical analysis is sufficiently rich at present to permit testing of empirical conclusions given any one of several kinds of raw data. Even such practical questions as designating optimal procedures for population exploitation and pest eradication are in principle answerable within the existing theoretical framework.

The great lack at the moment is sufficiently precise data. Until recently, precise conversion coefficients for evaluating census data in terms of energy have been missing. The error of mean estimates of field situations attributable to sampling variance in time and spaco remains enormous.

The few laboratory analyses that have been made do not disagree either with each other or with the results of field studies. Due to their paucity as well as to the high variance of field estimates the significance of this lack of contradiction is not evident.

We can conclude that ecological efficiency is probably of the order of magnitude of 5 to 20%. At present there is no evidence to indicate either a taxonomic, ecological or geographic variation in ecological efficiency. While evolution might tend towards a maximization of ecological efficiency, there is apparently no theoretical reason why any particular value should be preferred by all ecological systems. For any particular species, the percentage of consumed energy transmitted to predators cannot exceed the percentage assimilated. The assimilation percentage for most organisms at most times is of the order of 20–40% but this varies with the nature and abundance of the food supply.

At present, precise evaluation of efficiency in a few situations is likely to be more significant than first order approximations in many.

Acknowledgement

The research from my laboratory reported in this paper has been supported by a grant from the National Science Foundation (G 9058).

References

Armstrong, J. (1960). The dynamics of *Dugesia tigrina* populations and of *Daphnia pulex* populations as modified by immigration. 104 pp. Ph.D. dissertation. Department of Zoology, University of Michigan, Ann Arbor.

Beverton, R. and Holt, S. J. (1957). Fishery Investigations, Ser. II, Vol. XIX. London: Her Majesty's Stationery Office. "The Dynamics of Exploited Fish Populations".

Borecky, G. W. (1956). *Ecology* **37**, 719–727. Population density of the limnetic Cladocera of Pymatunign reservoir.

Clarke, G. (1946). *Ecol. Monogr.* **16**, 322–335. Dynamics of production in a marine area.

Comfort, A. (1961). *Sci. Amer.* **205**(2), 108–119. The life span of animals.

Dineen, C. G. (1953). *Amer. Midl. Nat.* **50**, 349–379. An ecological study of a Minnesota pond.

Englemann, M. D. (1961). *Ecol. Monogr.* **31**, 221–38. The role of soil arthropods in the energetics of an old field community.

Evans, F. C. and Lanham, U. (1960). *Science* **131**(3412), 1531–1532. Distortion of the pyramid of numbers in a grassland insect community.

Fisher, R. A. (1958). "The Genetical Theory of Natural Selection". New York: Dover Publications. xiv + 291 pages. (Oxford: Clarendon Press, 1930, x + 272 pages.)

Foster, C., Rapoport, C. A. and Trucco, E. (1957). *Gen. Sys.* **2**, 9–29. Some unsolved problems in the theory of non-isolated systems.

Frank, P. (1960). *Amer. Nat.* **94**(878), 357–372. Prediction of population growth form in *Daphnia pulex* cultures.

Golley, F. B. (1960). *Ecol. Monogr.* **30**(2), 187–206. Energy dynamics of a food chain of an old-field community.

Golley, F. B. (1961). *Ecology* **42**(3), 581–583. Energy values of ecological materials.

Hairston, N. G. (1959). *Ecology* **40**(3), 404–416. Species abundance and community organization.

Hairston, N. G., Smith, F. and Slobodkin, L. B. (1960). *Amer. Nat.* **94**(879), 421–425. Community structure, population control, and competition.

Hutchinson, G. E. (1959). *Amer. Nat.* **93**(870), 145–159. Homage to santa rosalia or why are there so many kinds of animals.

Ivlev, V. S. (1945). *Adv. mod. Biol., Moscow* **19**, 98–120. The biological productivity of waters. Trans. W. E. Ricker.

Ivlev, V. S. (1961). "Experimental Ecology of the Feeding of Fishes". Trans. Douglas Scott. New Haven, Yale University Press. viii + 302 pages.

Juday, C. (1940). *Ecology* **21**, 438–450. The annual energy budget of an inland lake.

Ketchum, B. H. and Redfield, A. C. (1949). *J. cell. comp. Physiol.* **33**, 281–300. Some physical and chemical characteristics of algae grown in mass culture.

Lindeman, R. L. (1942). *Ecology* **23**(4), 399–418. Trophic-dynamic aspect of ecology.

MacArthur, R. (1955). *Ecology* **36**(3), 533–536. Fluctuations of animal populations, and a measure of community stability.

MacArthur, R. (1960). *Amer. Nat.* **94**(874), 25–36. On the relative abundance of species.

MacArthur, R. and MacArthur, J. (1961). *Ecology* **42**(3), 594–597. On bird species diversity.

Margalef, R. (1958). Information theory in ecology. *Gen. Sys.* **3**, 36–71. Translated from *Mem. R. Acad. Barcelona*, **23**, 373–449 (1957).

Odum, E. P. and Odum, H. T. (1955). *Ecol. Monogr.* **25**, 291–320. Trophic structure and productivity of a windward coral reef community on Eniwetok atoll.

Odum, E. P. and Smalley, A. E. (1959). *Proc. Nat. Acad. Sci.* **45**(4), 617–622. Comparison of population energy flow of a herbivorous and a deposit-feeding invertebrate in a salt marsh ecosystem.

Odum, H. T. (1957). *Ecol. Monogr.* **27**, 55–112. Trophic structure and productivity of Silver Springs, Florida.

Odum, H. T. (1960). *Amer. Sci.* **48**(1), 1–8. Ecological potential and analogue circuits for the ecosystem.

Odum, H. T. and Pinkerton, R. (1955). *Amer. Sci.* **43**(2), 331–343. Time's speed regulator.

Patten, B. (1959). *Ecology* **40**(2), 221–231. An introduction to the cybernetics of the ecosystem trophic-dynamic aspect.

Pratt, D. M. (1943). *Biol. Bull.* **85**, 116–140. Analysis of population development in *Daphnia* at different temperatures.

Richman, S. (1958). *Ecol. Monogr.* **28**, 273–291. The transformation of energy by *Daphnia pulex*.

Ricker, W. E. (1946). *Ecol. Monogr.* **16**, 374–391. Production and utilization of fish populations.

Schrodinger, E. (1946). "What is Life?" viii + 91, New York: The Macmillan Co.

Slobodkin, L. B. (1954). *Ecol. Monogr.* **24**, 69–88. Population dynamics in *Daphnia obtusa Kurz*.

Slobodkin, L. B. (1959). *Ecology* **40**(2), 232–243. Energetics in *Daphnia pulex* populations.

Slobodkin, L. B. (1960). *Amer. Nat.* **94**(876), 213–236. Ecological energy relationships at the population level.

Slobodkin, L. B. (1961a). *Amer. Nat.* **95**(3), 147–153. Preliminary ideas for a predictive theory of ecology.

Slobodkin, L. B. (1961b). "Growth and Regulation of Animal Populations", 184 pp. Holt, Rinehart, and Winston Co, New York, New York.

Slobodkin, L. B. and Richman, S. (1961). *Nature, Lond.* **191**(4785), 299. Calories/gm in species of animals.

Teal, J. M. (1957). *Ecol. Monogr.* **23**, 41–78. Community metabolism in a temperate cold spring.

Watt, K. E. F. (1961). *Canad. Ent.* Supplement 19, **93**. Mathematical models for use in insect pest control.

Wiegert, R. (1962). (Personal communication.)

Quantitative Ecology and the Woodland Ecosystem Concept

J. D. OVINGTON

The Nature Conservancy, Monk's Wood Research Station, St. Ives, Huntingdonshire, England

I. Introduction

Trees, by virtue of their woody structure and longevity, normally attain a great size, and the forest combination of trees, shrubs and non-woody plants forms a multi-layered community frequently extending from the lowest soil depth to about 30 m above ground level in temperate forests and 55 m in tropical rain forests (Richards, 1952). Each plant layer supports a variety of animal life, has its own characteristic microclimate and is heterogeneous both vertically and horizontally. Of all terrestrial communities, woodlands are probably the most complex and massive, and not unexpectedly woodland ecologists have attempted to restrict their research to whatever features they felt were of greatest importance or could be recorded most readily.

Unfortunately these intensive but limited investigations have rarely been put into perspective against the general forest background and individual features have received undue emphasis as "indicators" of the overall woodland condition. The resultant lack of integration has caused an oversimplified approach to woodland ecology which has further hindered the effective co-ordination of results from different disciplines and areas. Multiple and more intensive use of forest land is inevitable as the world population multiplies, and the more thorough and comprehensive our knowledge of woodland ecology, the better will be the prospect for wise use and long-term conservation of the woodland resource. Some unifying concept embracing all aspects of woodland ecology is needed to bring forth a deeper understanding of woodland systems and to serve as the basis for their more rational utilization. It seems that the concept of ecosystem may provide the universal backcloth against which to show woodlands in all their patterned complexity.

II. The Ecosystem Concept

The term ecosystem was introduced by Tansley in 1935 to embrace not only living organisms and their remains, but also the whole complex of environmental factors, both biological and physical, operative in an ecological unit. Hills (1960) considers that so far as fundamental definitions are concerned, ecosystem is largely synonymous with biogeocoenose as coined by Sukachev (1944) in the U.S.S.R. and total site as used by Hills in Canada although there is some difference in emphasis. Various approaches to the study of forest ecosystems will undoubtedly arise because of geographic individuality and differences in interpretation, objectives and available knowledge, but this cannot affect the fundamental unifying nature of the concept (Rowe *et al.*, 1960). The potential use of the ecosystem concept is now becoming more fully recognized (Crocker, 1952; Sjörs, 1955), for by bringing together a variety of viewpoints it focuses attention upon the fundamental relationships and balance between woodland plants and animals, the physical environment of soil and climate, and the influence of man. Furthermore, it provides the key to a better understanding of the dynamic nature of woodlands not only throughout the year but on a long-term basis spanning generations of trees.

Ecosystems cannot be regarded as closed systems, isolated and independent, since interflow of energy and matter occurs between adjacent ecosystems in diverse and changing ways. The boundaries of ecological units may be delimited by differences in soil, climate or the distribution limits of plant or animal species, but since all of these are subject to change with time and are not necessarily precise nor coincident, it is unlikely that universally acceptable categories of ecosystems with discrete boundaries will ever be recognized. The flexibility of the ecosystem approach in relation to both spatial and time boundaries leads to some misunderstanding and criticisms of the concept because of its arbitrary nature and vagueness in this respect (Sukachev, 1960). In reality, the strength of the concept lies in this factor for, whilst it permits recognition of the considerable internal variation within ecological systems and the rarity of sharp boundaries between them, viz. the continuum (Curtis and McIntosh, 1951; Brown and Curtis, 1952) and the gradient analysis of Whittaker (1953, 1956), it does not exclude the practical classification of forest by the recognition of specified categories for particular purposes.

Evans (1956) has suggested that, since an ecological unit of any rank can be regarded as an ecosystem, the term can be interpreted to include the whole biosphere or have a more restricted nature, e.g. an acorn (Winston, 1956) or soil (Auerbach, 1958). Thus it is possible to envisage a hierarchy of increasingly inclusive ecosystems culminating in an all-encompassing world ecosystem (Rowe, 1961).

A. THE ECOSYSTEM CONCEPT IN RELATION TO FOREST RESEARCH

Fundamental research has not received so much attention in forestry as in horticulture or agriculture, partly because the forest has been regarded primarily as a wild crop, capable of sustaining itself indefinitely in spite of periodic logging. The advent of more intensive forestry, with the conversion of much forest land to agriculture and the increasing value of forest products, necessitates a greater emphasis on fundamental investigations and their application to sylvicultural practice. Peace (1961) has described some of the differences in approach to natural and artificial forests and stressed the importance of putting sylviculture on sound scientific grounds unfettered by preconceived notions.

Whilst the ecosystem concept has value in relation to a number of forestry problems, its greatest contribution to improving forestry practice probably will be that it provides a sound foundation for investigations designed to elucidate the functional processes of woodlands and to show the bearing of these processes on forest productivity and the possibilities for improving productivity on a long-term basis. Lindeman (1942) believed the concept of ecosystem to be of fundamental importance in interpreting the data of dynamic ecology. If forest research is to be fully effective, it needs to be orientated towards obtaining a better appreciation of ecosystem dynamics; particularly in relation to quantitative studies of the biological and physical processes affecting productivity and the accumulation, transformation and flow of energy and materials (water, mineral elements, etc.) through different woodland ecosystems. Our present knowledge of the quantitative aspects of the processes concerned in forest dynamics are unfortunately far too scanty and disjointed, yet these processes constitute an expression of the woodland ecosystem perhaps more significant than the trees themselves. Some of the processes that need to be considered in studies of woodland ecosystems are illustrated in Fig. 1.

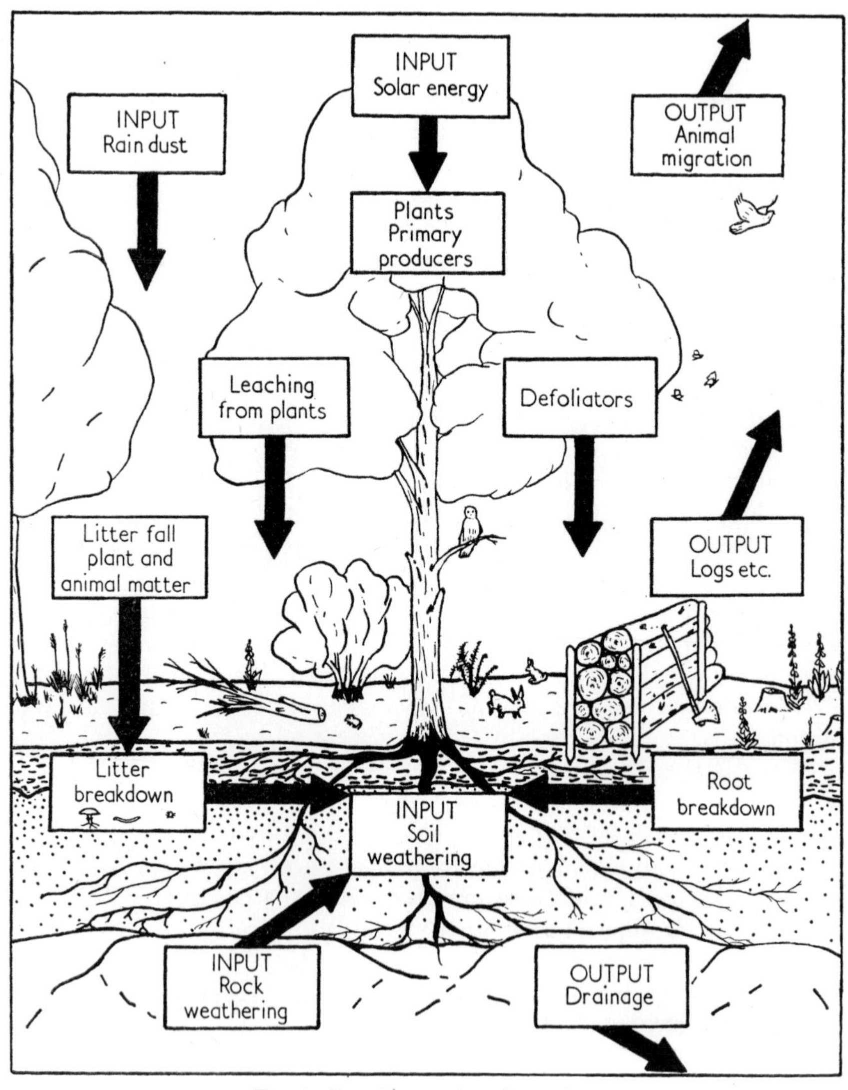

FIG. 1. Forest ecosystem dynamics.

B. TERMINOLOGY

Before discussing woodland ecosystem processes, it seems desirable to clarify the meaning of certain terms as used in this account for, in the past they have been given different meanings which have resulted in some misunderstanding. In this text, primary production refers solely to the production by the green plants, net primary production (apparent assimilation) being the total amount of organic matter synthesized by the plants as a result of photosynthesis and mineral absorption over a specified time, usually a year. Gross primary production includes net primary production plus the photosynthate used in respiration by the green plants. Biomass refers to the total quantity of organic matter present in the ecosystem at a stated time and may relate to particular organisms or groups of organisms. The term standing crop has been avoided since the forest crop can be extremely varied for it is not always restricted to timber but may also include items such as decorative foliage, Christmas trees, game, fruiting bodies of fungi and water. Unless otherwise stated all weights are given on an oven-dry basis.

III. Organic Matter Dynamics

One of the most obvious, yet frequently overlooked features of mature woodland ecosystems is that compared with other types of ecosystems they contain a relatively large mass of organic matter. The amount present is subject to change, reflecting the overall balance between organic matter production, harvesting and breakdown. Since energy flow, and mineral and water circulation are all closely linked to the organic system it seems appropriate to consider it first.

A. LACK OF WEIGHT DATA

Unfortunately, forestry, unlike agriculture, does not have a long history of crop weight recording, so there is no mass of detailed routine weight data for woodland ecosystems from which to show broad differences in weight on a geographical scale and to give a general indication of the magnitude of organic matter production. There are several reasons for this, the main one being that the value of a forest stand is normally assessed from estimates of the marketable volume of wood in the tree trunks. Wood-volume figures can be converted to dry weights since the specific gravities of different woods are known, but the results of such conversions have only a limited application to ecosystem studies because not all, or even complete, tree trunks are harvested. Further complications arise in that trees are harvested selectively over many years and tree boles form only part of the organic matter produced in woodland ecosystems. Not unnaturally most research data of tree bole size are expressed as volume rather than weight, to conform with the normal forest records and in some instances branches and leaves are also given on a volume basis. However sufficient weight data are available to enable some preliminary conclusions to be drawn and to serve as a guide for woodland ecologists, who are showing renewed interest in the determination of forest biomass.

B. ORGANIC MATTER IN WOODLAND ECOSYSTEMS

1. Plant Biomass

Normally much of the organic matter present in woodland ecosystems is plant material and in Table I plant biomass figures are given for several woodlands, selected primarily because the available records for them are relatively complete and show the overall distribution of the major types of plant material.

In addition to the results given in Table I, detailed data have been published relating to particular components of the organic system and these are broadly in accord with the values given in Table I. Burger (1929, 1935, 1945, 1947, 1948, 1950, 1951 and 1953) for instance, has determined leaf and branch weights in a large number of Swiss woodlands which vary in age from 13 to 92 years and contain different tree species. He obtained fresh weights, expressed as 10^3 kg per ha, which for leaves, ranged from 16 for Weymouth pine to 30 for Norway spruce, and for branches ranged from 36 for Scots pine to 61 for Norway spruce, the dry weight being about 40% of the fresh weight. Möller (1947) gives the following dry weights for leaves, expressed as 10^3 kg per ha: oak 2, ash 3, birch 2, Norway spruce 12, Scots pine 5 and larch 2; whilst Satoo *et al.* (1955, 1959) and Satoo and Senda (1958) give values of 3, 12 and 13 respectively for the tree leaves in stands of *Chamaecyparis obtusa*, *Zelkowa serrata*, and *Pinus densiflora* in Japan.

As stated earlier, there are few "weight per unit area of land" data available for tree trunks but tentative estimates of stem weights are given in Table II. These have been calculated from the stem-volume figures in standard yield tables for British and German woodlands on first quality sites and the average specific gravities for tree boles about 50 years old grown at a good site in England. Taking into account the average height, stocking density and age of the trees, as well as differences in site class, the estimated stem weights in Table II do not differ greatly from the values of Table I.

Shrubs are largely absent from most of the woodlands referred to in Table I and whilst this may be attributed to the fact that many of the woodlands are intensively managed and the tree crowns form a close canopy, the evidence from the more natural woodlands in the

Table does not suggest that the woodland shrub layer is typically very luxuriant. The dry weight of the herbaceous layer varies considerably, the understorey vegetation being absent in dense woodlands and very luxuriant in open sheltered woodlands where it may exceed the tree leaves in weight. Pase and Hurd (1957) found that the biomass of the air-dry herbage increased from 25 to 1 857 kg per ha when the basal area of the trees decreased from 50 m^2 per ha to 0 on clear-cut areas. From evidence obtained with trenched plots, Ellison and Houston (1958) suggest that root competition between the trees and understorey plants is an important factor, perhaps more important than shading, in limiting the growth of the understorey vegetation. Tamm (1953) found that in an open spruce forest with a mossy ground flora, mainly of *Hylocomium splendens*, the living moss biomass was about 2 080 kg per ha. The annual production of about 1 298 kg per ha by the moss was closely related to the light supply when this was low, and at high light intensities to the nutrient supply contained in the rain-water dripping from the tree canopies.

Numerous weight determinations have been reported for the layer of organic matter formed over the mineral soil; in temperate regions little organic matter accumulates under most hardwood tree species, beech can be a notable exception, whilst in coniferous stands a large weight of surface organic matter may persist. Metz (1954) comparing hardwood and pine stands in South Carolina, U.S.A., found the oven-dry weight of organic matter, calculated as 10^3 kg per ha, ranged from 6 to 12 in hardwood stands, from 13 to 18 in mixed hardwood and pine stands, and from 17 to 23 in pine stands. Further north, greater amounts of organic matter tend to accumulate over the mineral soil. Alway and various associates (Alway, 1930; Alway and Harmer, 1927; Alway and Rost, 1927) give a range of 18 to 76×10^3 kg per ha for the weight of surface organic matter in nine virgin forest stands in Minnesota, U.S.A., 16 to 58, for 21 forests in north eastern Minnesota, and 18 to 38 for five pine stands in North Minnesota. Ovington (1954a) in England found the weight of the surface organic matter varied between 4 to 6×10^3 kg per ha in nine hardwood stands and from 7 to 35 in 19 plantations of coniferous trees. The 33-year-old stand of Scots pine in Table I has an unusually huge amount of surface organic matter, but this plantation is particularly dense, has been neglected and never thinned so that the soil organic layer contains a large proportion of fallen trees and branches. In this respect it closely resembles the natural Douglas fir stands, the differences in weight of the organic matter of the forest floor in managed and natural stands being well-shown by the Douglas fir series. No biomass figures for natural populations of forest fungi and bacteria are known but bacteria and fungi are included in the organic

TABLE I. *Plant Biomass*

Oven-dry weight 10^3 kg per ha

Trees	*Pinus densiflora*	*Pinus densiflora*	*Pinus sylvestris*	*Pinus sylvestris*	*Pinus sylvestris*	*Pinus sylvestris*	*Pinus sylvestris*	*Pinus strobus*	*Pinus strobus*	*Pinus nigra*	*Pinus nigra*
Location	Tiba Japan	Tiba Japan	East England	South Scotland	West England	East England	Northeast Scotland	Hokkaido Japan	Hokkaido Japan	West England	Northeast Scotland
Status	Plantation	Plantation	Plantation	Plantation	Plantation	Plantation	Plantation	Plantation	Plantation	Plantation	Plantation
Stand Number	1	2	3	4	5	6	7	8	9	10	11
Ages of trees in years	16	16	23	33	47	55	64	41	41	46	48
Tree height in metres	—	—	8	14	23	16	16	—	—	23	14
Number of trees per hectare	—	—	3 640	4 260	445	760	815	—	—	482	1 112
Living plants											
Tree layer											
Whole canopy	16·4	12·9	19·3	21·3	27·0	20·0	21·4	26·0	24·2	30·2	16·8
Fruit	0	0	0·4	—	—	0·5	—	—	—	—	—
Leaves	5·4	5·3	5·1	7·3	—	7·2	4·7	7·4	10·1	—	5·6
Branches	11·0	7·6	13·8	14·0	—	12·3	16·7	18·6	14·1	—	11·2
Trunks	36·6	85·5	44·3	118·8	129·6	96·7	97·2	78·7	180·0	212·0	95·1
Shrub layer											
Above ground	0	0	0	0	0	0	0	—	—	0	0
Herb layer											
Above ground	6·2	6·3	0·2	<0·1	7·0	2·6	—	—	—	6·8	—
Roots	—	—	28·1	36·1	—	34·1	—	—	—	—	—
Dead plant material											
On trees	—	—	13·0	19·3	—	10·0	—	—	—	—	—
Litter on ground (L, F and H)	—	—	29·8	110·6	12·7	45·0	—	—	—	21·7	—
Removed as harvested trunks	—	—	15·1	0	213·3	166·9	—	—	—	184·0	—
Reference	A	A	B	C	D	B	E	F	F	D	E

Key to References shown in Table I. A. Satoo *et al.* 1955; B. Ovington, 1957a, 1959a; C. Ovington and Madgwick, 1959a; D. Ovington, 1954a, 1955, 1956a, b, c; E. Wright and Will, 1958; F. Senda and Satoo, 1956; G. Sonn, 1960; H. Satoo, unpublished data; I. Tamm and Carbonnier, 1961; J. Heilman, 1961; K. Smirnova and Gorodentseva, 1958; L. Ovington and Madgwick, 1959b; M. Tamm, in manuscript; N. Ovington and Heitkamp, in manuscript; O. Möller *et al.*, 1954; P. Miller, in manuscript; Q. Satoo *et al.*, 1956; R. Greenland and Kowal, 1960; S. Ogawa *et al.*, 1961.

TABLE I.—*continued*

Trees	*Picea abies*	*Picea abies*	*Picea abies*	*Picea abies*	*Picea abies*	*Picea abies*	*Picea abies*	*Picea abies*	*Picea abies*	*Picea abies*	*Picea abies*	*Picea abies*
Location	South England	U.S.S.R.	U.S.S.R.	Titibu Japan	Hokkaido Japan	Hokkaido Japan	West England	West England	Sweden	Sweden	U.S.S.R.	U.S.S.R.
Status	Plantation	—	—	Plantation	Plantation	Plantation	Plantation	Plantation	Plantation	Natural	—	—
Stand Number	12	13	14	15	16	17	18	19	20	21	22	23
Age of trees in years	20	24	38	39	46	46	47	47	52	58	60	93
Tree height in metres	11	—	—	—	—	—	21	21	17	17	—	—
Number of trees per hectare	6 365	—	—	—	—	—	667	937	1 125	924	—	—
Living plants												
Tree layer												
Whole canopy	61·1	—	—	30·4	23·6	27·2	31·9	80·3	26·4	23·4	—	—
Fruit	—	—	—	0	—	—	—	—	—	—	—	—
Leaves	—	3·0	9·6	24·6	16·9	18·6	—	—	10·8	9·1	11·1	10·0
Branches	—			5·8	6·7	8·6	—	—	15·6	14·3	195·6	249·5
Trunks	157·2	69·1	113·1	138·3	63·2	140·2	107·9	182·4	105·3	85·2		
Shrub layer												
Above ground	0	8·9	—	—	0·2	0·2	0	0	—	—	—	—
Herb layer												
Above ground	0	—	—	0	1·2	1·6	0	0	—	—	—	1·0
Roots	—	20·2	38·1	—	—	—	—	—	—	—	64·7	65·4
Dead plant material												
On trees (branches)	—	—	—	—	—	—	—	—	3·3	2·6	—	—
Litter on ground (L, F and H)	17·4	—	—	—	—	—	25·0	26·0	—	78·0	—	—
Removed as harvested trunks	6·6	—	—	—	—	—	210·5	178·7	—	—	—	—
Reference	D	G	G	H	H	H	D	D	I	I	G	G

TABLE I. — *continued*

Trees	*Pseudotsuga taxifolia*	*Pseudotsuga taxifolia*	*Pseudotsuga taxifolia*	*Pseudotsuga taxifolia*	*Pseudotsuga taxifolia*	*Pseudotsuga taxifolia*	*Pseudotsuga taxifolia*	*Pseudotsuga taxifolia*	*Larix decidua*
Location	South England	East England	Washington U.S.A.	Washington U.S.A.	Washington U.S.A.	Washington U.S.A.	West England	Washington U.S.A.	West England
Status	Plantation	Plantation	Natural	Natural	Natural	Natural	Plantation	Natural	Plantation
Stand Number	24	25	26	27	28	29	30	31	32
Age of trees in years	21	22	30	32	38	38	47	52	46
Tree height in metres	13	11	9	9	14	17	29	17	17
Number of trees per hectare	1 700	2 100	1 151	1 636	1 151	648	297	1 157	420
Living plants									
Tree layer									
Whole canopy	24·5	82·7	14·3	9·9	16·5	22·9	49·7	29·9	43·6
Fruit	—	—	—	—	—	—	—	—	—
Leaves	—	—	8·0	5·3	8·0	9·0	—	12·0	—
Branches	—	—	6·3	4·6	8·5	13·9	—	17·9	—
Trunks	90·3	92·1	22·9	23·3	69·8	130·0	202·7	174·8	145·8
Shrub layer									
Above ground	0	0					0		0
Herb layer			11·0	3·2	1·3	1·8		0·1	
Above ground	0·2	0					3·6		4·6
Roots	—	—	25·1	20·7	10·0	16·9	—	12·3	—
Dead plant material									
On trees (branches)	—	—	1·3	2·6	10·2	6·7	—	11·2	—
Litter on ground (L, F and H)	10·9	21·7	61·4	35·0	63·0	46·9	8·3	117·3	34·8
Removed as harvested trunks	36·5	21·9	—	—	—	—	206·9	—	85·6
Reference	D	D	J	J	J	J	D	J	D

Table I. — *continued*

Trees	*Betula verrucosa*	*Betula verrucosa*	*Betula verrucosa*	*Betula verrucosa*	*Betula verrucosa*	*Betula verrucosa*	*Betula verrucosa*	*Betula maximowicziana*	*Betula maximowicziana*	*Betula maximowicziana*	*Alnus incana*
Location	Moscow U.S.S.R.	East England	Central England	North Sweden	Moscow U.S.S.R.	Central England	Moscow U.S.S.R.	Hokkaido Japan	Hokkaido Japan	Hokkaido Japan	East England
Status	Natural	Plantation	Natural	Natural	Natural	Natural	Natural	Natural	Natural	Natural	Planted
Stand Number	33	34	35	36	37	38	39	40	41	42	43
Age of trees in years	20	22	24	25	40	55	67	47	47	47	22
Tree height in metres	11	—	9	12	19	18	26	—	—	—	12
Number of trees per hectare	—	4 028	4 990	2 350	—	880	—	—	—	—	1 656
Living plants											
Tree layer											
Whole canopy	15·1	17·7	14·2	8·5	16·0	29·5	14·1	12·9	14·7	17·0	28·0
Fruit	—	—	—	—	—	—	—	0	0	0	—
Leaves	3·8	—	2·4	2·7	3·3	2·5	2·8	1·8	2·6	2·2	—
Branches	11·3	—	11·8	5·8	12·7	27·0	11·3	11·1	12·1	14·8	—
Trunks	45·7	43·1	48·0	38·9	190·7	134·5	156·7	77·7	128·3	100·0	83·1
Shrub layer											
Above ground	—	0	0	—	—	0	—	0	0	0	0
Herb layer											
Above ground	—	2·2	—	—	—	—	—	—	—	—	2·1
Roots	19·5	—	16·9	—	40·9	49·8	43·1	—	—	—	—
Dead plant material											
On trees (branches)	—	—	2·7	—	—	1·7	—	—	—	—	—
Litter on ground (L, F and H)	—	4·7	—	—	—	—	—	—	—	—	5·4
Removed as harvested trunks	—	0	0	—	—	0	—	—	—	—	13·9
Reference	K	D	L	M	K	L	K	H	H	H	D

Table I. — *continued*

Trees	*Quercus petraea*	*Quercus*	*Quercus*	*Quercus robur*	*Quercus*	*Quercus borealis*	*Quercus*	*Fagus sylvatica*	*Fagus sylvatica*	*Fagus sylvatica*	*Nothofagus obliqua*	*Nothofagus truncata*
Location	South England	U.S.S.R.	U.S.S.R.	West England	U.S.S.R.	Minnesota U.S.A.	U.S.S.R.	West England	Denmark	Denmark	South England	N. Island N. Zealand
Status	Plantation			Plantation		Natural		Plantation	Plantation	Plantation	Plantation	Natural
Stand Number	44	45	46	47	48	49	50	51	52	53	54	55
Age of trees in years	21	22	42	47	56	57	200	39	46	85	22	110
Tree height in metres	5	—	—	17	—	17	—	17	18	26	9	21
Number of trees per hectare	10 102	—	—	618	—	800	—	815	3 110	320	3 558	490
Living plants												
Tree layer												
Whole canopy	14·1	—	—	21·7	—	53·0	—	35·5	—	—	32·0	44·7
Fruit	—	—	—	—	—	0·1	—	—	—	—	—	—
Leaves	—	2·1	3·2	—	3·8	3·5	3·2	—	2·7	2·7	—	2·7
Branches	—			—		49·5		—			—	42·0
Trunks	28·3	53·6 (branches and trunks)	136·1 (branches and trunks)	106·6	188·4 (branches and trunks)	111·9	400·1 (branches and trunks)	97·9	131·1 (branches and trunks)	232·7 (branches and trunks)	48·7	224·8
Shrub layer												
Above ground	0	5·6	1·2	0	1·2	0·5	3·5	0	—	—	0	—
Herb layer												
Above ground	0·6	0·2	0·3	1·6	0·4	0·1	0·2	0·2	—	—	0·8	—
Roots	—	28·7	29·0	—	38·3	15·0	42·7	—	26·2	46·3	—	39·2
Dead plant material												
On trees (branches)	—	—	—	—	—	21·9	—	—	—	—	—	1·1
Litter on ground (L, F and H)	6·0	—	—	3·7	—	36·7	—	10·7	—	—	4·0	16·7
Removed as harvested trunks	0	—	—	56·8	—	—	—	63·6	—	—	14·1	—
Reference	D	G	G	D	G	N	G	D	O	O	D	P

TABLE I. — *continued*

Trees	*Castanea sativa*	*Populus davidiana*	*Cinnamomum camphora*	Tropical evergreen	Tropical deciduous	Mixed Dipterocarp	Mixed Savanna	Temperate evergreen	Evergreen gallery
Location	West England	Hokkaido Japan	Tiba Japan	Belgian Congo	Ghana	Thailand	Thailand	Thailand	Thailand
Status	Plantation	Natural	Plantation	Natural	Natural	Natural	Natural	Natural	Natural
Stand Number	56	57	58	59	60	61	62	63	64
Age of trees in years	47	40	48	18	50	—	—	—	—
Tree height in metres	21	—	—	—	61	25	25	25	29
Number of trees per hectare	427	—	—	—	6 252	1 576	1 340	2 933	16 200
Living plants									
Tree layer									
Whole canopy	8·2	10·2	22·3	—		—	—	—	—
Fruit	—	0	0	—		—	—	—	—
Leaves	—	2·2	4·1	6·5		5·0	5·0	14·8	19·8
Branches	—	8·0	18·2						
Trunks	108·4	104·8	192·4	116·4	261·6	45·4	53·4	166·0	275·2
Shrub layer									
Above ground	0	—	—	—		0	0·4	0·2	0·2
Herb layer									
Above ground	1·2	3·6	—	—		0·7	0·8	0	0
Roots	—	—	—	31·3	24·7	15·8	18·6	54·3	88·5
Dead plant material									
On trees (branches)	—	—	—	17·3	71·7	—	—	—	—
Litter on ground (L, F and H)	4·1	—	—	5·6	2·3	3·7	2·8	62·6	3·0
Removed as harvested trunks	59·5	—	—	—	—	—	—	—	—
Reference	D	Q	H	R	R	S	S	S	S

TABLE II. *Tree Trunks at Quality Class I Sites*

Tree species	Country	Standing tree trunks				Tree trunks as thinnings	Source of volume figures for conversion to oven-dry weights
		Age in years	Height in m	Number per ha	Oven-dry weight 10^3 kg/ha	Cumulative weight 10^3 kg/ha	
Pinus sylvestris	Britain	50	20	556	169	154	Hiley, 1954
Pinus sylvestris	Germany	50	19	998	141	54	Schwappach, 1912
Pinus sylvestris	Britain	75	27	297	230	271	Hiley, 1954
Pinus sylvestris	Germany	140	32	223	229	261	Schwappach, 1912
Pinus nigra	Britain	50	24	195	205	200	Hiley, 1954
Picea abies	Britain	50	23	605	219	157	Hiley, 1954
Picea abies	Germany	50	21	1 468	168	41	Schwappach, 1912
Picea abies	Germany	120	36	284	307	328	Schwappach, 1912
Pseudotsuga taxifolia	Britain	50	33	309	305	210	Hiley, 1954
Abies grandis	Britain	50	37	235	285	317	Christie and Lewis, 1961
Abies alba	Germany	50	19	1 800	182	54	Schwappach, 1912
Abies alba	Germany	120	34	400	461	276	Schwappach, 1912
Larix decidua	Britain	50	24	309	135	123	Hiley, 1954
Quercus robur	Britain	50	18	667	99	64	Waters and Christie, 1958
Quercus robur	Germany	50	18	971	103	75	Schwappach, 1912
Quercus robur	Britain	150	31	62	163	297	Waters and Christie, 1958
Quercus robur	Germany	200	34	100	354	426	Schwappach, 1912
Fagus sylvatica	Britain	50	21	642	135	108	Waters and Christie, 1958
Fagus sylvatica	Germany	50	19	—	152	46	Schwappach, 1912
Fagus sylvatica	Germany	140	38	—	413	291	Schwappach, 1912
Fagus sylvatica	Britain	150	33	79	295	443	Waters and Christie, 1958
Betula verrucosa	Germany	50	21	553	89	37	Schwappach, 1912
Betula verrucosa	Germany	80	26	230	117	97	Schwappach, 1912

material overlying the mineral soil, since the microflora cannot be separated easily. Höfler (1937) collected fruiting bodies of fungi in beech-oak forests and found a wide range of weights, up to $0{\cdot}18 \times 10^3$ kg per ha. Russell (1956) estimates the oven-dry weight of bacteria and fungi in Rothamsted arable soil to be approximately 0·34–0·78 and $0{\cdot}34 \times 10^3$ kg per ha respectively.

Relatively few records of root weights are available for forest ecosystems and although the published data give similar values on the whole, experience in root sampling indicates that these may be seriously in error for it is difficult to ensure that all roots are collected and no mineral soil is incorporated within the root mass. Frequently, the weight of the root system exceeds that of the tree canopy. Significant amounts of organic matter are present in the mineral soil but since it is difficult to distinguish between that due to plant roots and that due to other organic matter — e.g. leaves, etc. incorporated from the litter, or material such as fungal spores (Burges, 1950), pollen grains (Dimbleby, 1961) and the soil fauna — no figures other than root weights are given for the organic matter in the mineral soil.

The proportionate distribution of plant organic matter within woodland ecosystems is influenced by numerous factors such as the species of trees present and intensity of forest management, and consequently varies greatly. In the middle-aged, well-grown woodlands listed in Table I about 75% of the total plant biomass is contained within the living trees, the amount in the tree trunks exceeding that of the roots and canopy.

If the weights of the different types of plant matter are added together for the individual woodlands of Table I and appropriate adjustments made for missing components, it is noteworthy that the total oven-dry weights of the plant biomass in the better grown, close canopied stands is about 300 to 350×10^3 kg per ha. Ogawa *et al.* (1961) have made some tentative calculations of plant biomass for two luxuriant forests of tropical Africa described by Aubréville (1938) and Jones (1956) and obtained weights of 200 and 227×10^3 kg per ha exclusive of roots. They suggest that throughout different latitudes the plant mass in well-grown forests is fairly uniform in dry weight ranging between 200 and 350×10^3 kg per ha irrespective of climatic zone. However, this apparent "uniformity" of dry weight may be fortuitous, resulting from the limited number of woodlands sampled and certainly the sequence of forest types observed in passing northwards through Scandinavia for example, gives the impression of a progressive decrease in plant biomass. Their estimates of plant biomass weight for tropical forests may be low since Greenland and Kowal (1960) found the total weight of the vegetation cover in a disturbed moist tropical forest in

Ghana to be about 360×10^3 kg per ha. Furthermore, it can be seen from Table II that the weight of tree boles alone in some old forests on good quality sites in Great Britain and Germany would be expected to exceed 350×10^3 kg per ha. On a more restricted regional basis there may be some degree of uniformity in the maximum amount of plant biomass that different forests attain on the better sites. Regionally, the trees of the characteristic dominant species tend to be approximately the same size when mature and there are various compensating mechanisms in forest ecosystems tending towards uniformity of biomass weight. For example, well-stocked, immature woodlands have a large number of small trees but, as the trees grow in stature, mortality occurs and the older woodlands have a small number of large trees. Similarly there is a compensating mechanism between the plant layers, thus as the overstorey becomes sparser the understorey vegetation becomes more luxuriant. If there is a regional plant carrying capacity for woodland ecosystems then this could be equal for different regions, but it seems doubtful that it is universally constant irrespective of climatic conditions. Figures of the plant biomass of old, natural redwood or Douglas fir forests of western U.S.A., of *Eucalyptus* forests in Australia and of dense tropical forests, would be very interesting to compare because these, of all forest ecosystems, probably contain the greatest mass of plant organic matter. Particularly in the undisturbed N.W. American forests, there is a tendency for a huge mass of organic matter to accumulate on the ground, since after falling, the large tree trunks decompose relatively slowly. This would give a very different distribution of plant biomass from that of the intensively managed woodlands of Western Europe.

2. *Animal Biomass*

The problem of determining the biomass of forest animals is much more formidable than for plants, if we exclude fungi and bacteria from the flora, because many more animal than plant species are present; frequently identification of animal species is difficult, animals are relatively mobile and the fauna is subject to extreme fluctuations in numbers both throughout the year and from year to year (Macfadyen, 1957). Furthermore, some animals which may be of considerable ecological significance in woodland dynamics are only present at certain times of the year. Others use the forest covered areas primarily for cover and feed largely on the vegetation in openings, track-ways, roads or neighbouring fields. The extent to which they should be included in the forest biomass is questionable.

Insufficient data are available to provide a reasonably complete picture of the total weight of the fauna in different forests and to show the weight distribution within the various woodland strata but from a comparison of Tables I and III it seems that the total biomass of the fauna is likely to be much less than that of the plants. However, animal biomass figures give little indication of the functional activity of the organisms within the ecosystem and are probably of much more limited value than plant biomass figures in relation to ecosystem dynamics. Nevertheless their determination is a necessary first stage in evaluating the quantitative role of the fauna in forest processes.

3. *Changes in Biomass*

Changes in the weight of organic matter in woodland ecosystems occur both during the year and on a longer term basis. In relatively stable, mature, natural forests the weight of plant organic matter present is presumably fairly constant from year to year, there being an overall equilibrium between production and decomposition, although an intricate patchwork of accretion or loss may exist as old trees die and are replaced by new ones. In present-day woodlands this situation occurs only rarely, for in addition to natural events such as windblow or fire, the widespread effects of man have upset any natural equilibrium over large areas.

When new woodlands become established a continuous build up of organic matter takes place, and this can be seen best where new forests are successfully planted on areas deficient in organic matter, such as sand dunes or abandoned fields. The rate at which plant organic matter accumulates depends upon many factors but the highly productive, intensively managed woodlands of Table I contain approximately 350×10^3 kg per ha of plant organic material when about 50 years old giving an average annual rate of accretion of about 7×10^3 kg per ha. In some years the annual build up of plant organic matter greatly exceeds the average figure which includes the early stages when true woodland conditions do not exist and the older stages when large amounts of timber are being harvested.

Current annual changes in the weights of trees, ground flora and the organic matter over the surface of the mineral soil in plantations of *Pinus sylvestris* growing at a good pine site in the English Breckland are illustrated in Fig. 2 (Ovington, 1957a, 1959b). When the pine plantations are at the dense pole stage, i.e. from 12 to 20 years of age, there is a rapid build up of organic matter in the trees and on the forest floor whilst the ground vegetation becomes sparser, until it is only present in a few small patches. Afterwards, with the advent of thinning, gaps are created in the tree canopy, the ground vegetation begins to spread and increase in weight and, although organic matter continues to accumulate on the forest floor, it does so at a slower rate. The rate

TABLE III. *Animal Biomass*

Organism	Woodland	Location	Biomass 10^3 kg per ha	Comments		Source of data
Tree canopy invertebrates excluding Acarina and Collembola	Plantations Scots Pine	England	0·00010–0·00500	Oven-dry weight	Period September to March	A
Tree canopy invertebrates excluding Acarina and Collembola	Plantations Corsican Pine	England	0·00001–0·00200	Oven-dry weight	Period September to March	A
Insects — mainly beetles, leafhoppers, delphacids and dipterans	Natural Willow	U.S.A. Tennessee	0·00088	Fresh weight		B
Parus caeruleus (Blue tit)	Plantations of Scots and Corsican Pines	England	0·00100–0·02200	Oven-dry weight	Period September to April	C
Parus ater (Coal tit)	Plantations of Scots and Corsican Pines	England	0·00200–0·06200	Oven-dry weight	Period September to April	C
Regulus regulus (Gold crest)	Plantations of Scots and Corsican Pines	England	0·00700–0·01800	Oven-dry weight	Period September to April	C
Total bird population	Spruce	Czechoslovakia	0·00048	Fresh weight		D
Total bird population	Beech-maple-fir-spruce	Czechoslovakia	0·00117	Fresh weight		D
Total bird population	Oak-hornbeam	Czechoslovakia	0·00115	Fresh weight		D
Apodemus sylvaticus (Wood mouse)	Hardwoods in general	England	0·00025–0·00198	Fresh weight		E
Clethrionomys glareolus (Bank vole)	Hardwoods in general	England	0·00025–0·00198	Fresh weight		E
Sorex araneus (Common shrew)	Hardwoods in general	England	0·00012–0·00099	Fresh weight		E
Talpa europaea (Mole)	Hardwoods in general	England	0·00099–0·00494	Fresh weight		E
Oryctolagus cuniculus (Rabbit)	Hardwoods in general	England	0·00321–0·01284	Fresh weight		E
Sciurus carolinensis (Grey squirrel)	Hardwoods in general	England	0·00068–0·00679	Fresh weight		E
Dama dama (Fallow deer)	Hardwoods in general	England	0·00212	Fresh weight		E
Vulpes vulpes (Fox)	Hardwoods in general	England	0·00009–0·00023	Fresh weight		E
Meles meles (Badger)	Hardwoods in general	England	0·00014	Fresh weight		E
Mustela erminea (Stoat)	Hardwoods in general	England	0·00006	Fresh weight		E
Mustela nivalis (Weasel)	Hardwoods in general	England	0·00005	Fresh weight		E

TABLE III. — *continued*

Organisms	Woodland	Location	Biomass 10^3 kg per ha	Comments		Source of data
Microarthropoda	Conifer plantations	Denmark	0·11000	Fresh weight	Soil only	F
Nematoda	Conifer plantations	Denmark	0·04000–0·05000	Fresh weight	Soil only	F
Enchytraeidae	Conifer plantations	Denmark	0·03000–0·25000	Fresh weight	Soil only	F
Acarina and Collembola	Beech	Denmark	0·70770	Fresh weight	Soil only, mull	G
Acarina and Collembola	Beech	Denmark	0·16570	Fresh weight	Soil only, mor	G
Acarina and Collembola	Spruce	Denmark	0·09840	Fresh weight	Soil only, mor	G
Enchytraeidae	Douglas fir plantation	Wales	0·02700–0·13200	Fresh weight	Soil only, mor	H

A. Gibb, 1960 B. Crossley and Howden, 1961 C. Grimshaw *et al.*, 1958
D. Turček, 1956 E. Southern — private communication F. Nielsen, 1955
G. Bornebusch, 1930 H. O'Connor, 1957

of build up of organic matter into the standing trees remains high up to about 35 years, after which it declines until at 45 years of age new growth and harvesting are more or less in balance. Although fairly heavy thinning has taken place and large amounts of timber have been removed, the mean annual increment of plant organic matter into the ecosystem as a whole for the first 55 years after planting is about 5×10^3 kg per ha, whilst the maximum current annual increment recorded is 9×10^3 kg per ha, attained around 20 years of age.

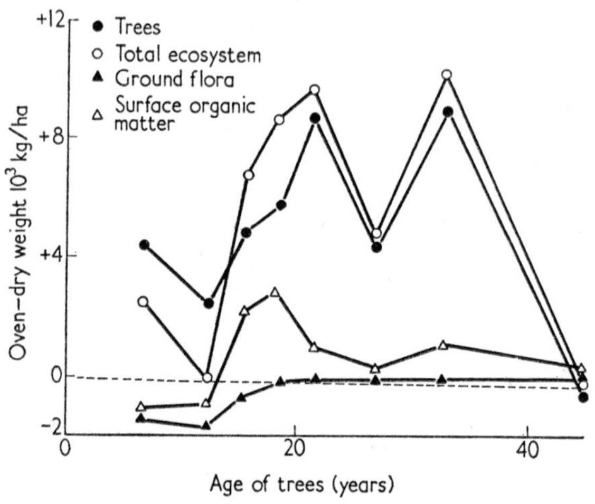

FIG. 2. Current annual change in the amount of organic matter present in plantations of *Pinus sylvestris*.

In some woodland ecosystems, the build up of organic matter may take place more rapidly than in the Scots pine plantations. In Britain, organic matter generally accumulates more rapidly in coniferous, evergreen woodlands than in those of deciduous, angiosperm trees. For example, the total dry weights of the above ground parts of the trees, ground flora and the dead organic matter over the mineral soil in neighbouring plantations of Douglas fir, Scots pine and oak, all 47 years old and growing under identical conditions, are 264, 177 and 121×10^3 kg per ha respectively (Ovington, 1954a, 1955, 1956a). The smaller weight of organic matter in the oak woodland cannot be attributed to overthinning and, in fact, the amount of timber harvested from the oak plot is about a quarter of that from either of the coniferous stands. The rate of accretion of organic matter and the number of years needed for maximum accretion to be achieved vary considerably for different woodlands depending upon site conditions, intensity of management and the species of tree present.

Apart from instances when members of the fauna cause serious damage to the plant cover, progressive long term changes in the biomass of the animal organic matter are poorly documented but there seems little doubt that they do occur. Lack (1933, 1939) for example, has recorded the progressive changes in the avifauna following afforestation in England, and it appears likely that these changes in species of bird present result in a change of avian biomass.

C. ORGANIC MATTER PRODUCTION

1. Net Primary Production

i. Determination of Net Primary Productivity. Biomass is an unsatisfactory measure of net primary productivity as shown by Table IV; in the case of perennials, such as trees, the discrepancy between the two values increases greatly with age. Because of the immensity of woodland ecosystems and the heterogeneity of woodland sites, the comprehensive sampling required to determine net primary productivity presents an Herculean task. Various techniques based on some form of selective harvesting have been developed to reduce the amount of sampling but there is still considerable scope for further improvement. However, fairly intensive sampling is inevitable since the woodland plants capable of photosynthesis exhibit a wide range of form and life history, and live under very different conditions of microclimate. Ideally, net primary productivity should be determined separately for individuals of all the plant species present (Odum, 1960) and account taken of differences between individuals of the same species. Significant differences in productivity are known to occur between trees depending upon their status within the tree canopy. Burger (1953) reports that, on average, a 65-year-old plantation of Norway spruce needs 2 300 kg of fresh leaves to produce a cubic metre of wood but the dominant trees of the stand only require 1 300 kg of leaves.

TABLE IV

Biomass and Net Primary Productivity of Trees of Pinus sylvestris *from the Time of Planting*

	Oven-dry weight 10^3 kg per ha									
Age of plantations in years	3	7	11	14	17	20	23	31	35	55
Biomass of living trees (*a*)	<0·1	7·5	25·9	31·9	45·6	58·8	78·7	113·8	146·1	140·7
Net primary productivity (*b*)	<0·1	8·1	30·9	50·0	85·4	118·9	217·7	348·4	445·2	682·9
(*b*) : (*a*)	1	1·1	1·2	1·6	1·9	2·0	2·8	3·1	3·1	4·9

A serious source of error in determining net primary productivity arises through too infrequent sampling so that the plants are not harvested at their maximum weights or some plant parts such as bud scales of flowers are missed completely. For example, the annual production of inflorescences in stands of aspen, *Populus tremuloides*, and of male cones in stands of white pine, *Pinus strobus*, amounts to about 230 and 656 kg per ha, equal to 5 and 25% respectively of the annual production of leaves; if sampling is restricted to summer and autumn these would not be included in primary production estimates. Estimates of net primary productivity as root material may be very inaccurate since they are normally calculated from differences in root biomass from year to year but it is known that, just as for the tree leaves, there is an annual mortality of the smaller roots. According to Orlov (1955) this amounts to about 5% of the total root production. Another source of error arises through failure to take into account the amount of primary production eaten by animals. The fact that plants may show little evidence of animal cropping at the time of sampling can be misleading; for instance *Tortrix viridana* frequently completely defoliates oak trees in early spring, which later grow a new set of leaves so that leaf production is about twice the weight of leaves recorded by a single summer sampling. On the whole, errors such as these all lead to an underestimation of net primary productivity and the published figures are likely to be too small.

ii. Change of Net Primary Productivity with Age. The annual net primary productivities for the trees and for the ecosystem as a whole (trees plus ground flora) at different ages in the series of plantations of *Pinus sylvestris* are shown in Fig. 3, the estimates being subject to some of the errors described previously. The mean annual productivity of the trees increases steadily to just over 12×10^3 kg per ha at about 35 years of age, after which productivity decreases slightly. The curve of current annual productivity for the trees is more irregular than the mean curve but when smoothed out shows a relatively rapid increase in productivity to a maximum value of just over 22×10^3 kg per ha, which is achieved by about 20 years of age and is maintained for several years before declining fairly rapidly after 35 years.

In the youngest plantations the trees are small and not contiguous, whilst in the older plantations thinning is repeatedly creating gaps in the forest so that the tree canopies and root systems do not continually occupy the available crown and root space. Maximum current annual productivity by the ecosystem is therefore achieved when the dominance of the trees, as expressed by canopy and root development is greatest, so that it is not surprising that the biomass of the understorey vegetation is at a minimum at this stage. When the plantations are 20 years

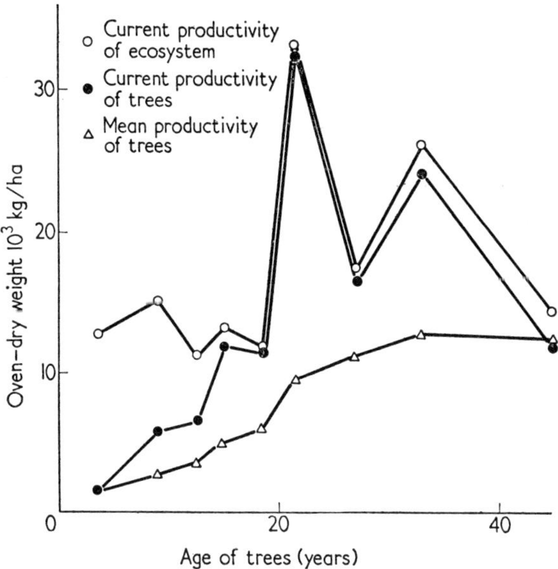

FIG. 3. Annual net production of organic matter by the trees and ecosystem (trees + understorey vegetation) in plantations of *Pinus sylvestris*.

of age and the current productivity of the trees is high, the above ground parts of the understorey vegetation only amount to about $0{\cdot}2 \times 10^3$ kg per ha and the annual contribution of the ground flora to the current net primary productivity of the ecosystem is negligible compared with that of the trees. The contribution of the ground flora to current productivity of the ecosystem is greatest at the juvenile and old stages of plantation development, but the combined productivities of trees and ground flora at these stages only gives an average value for current net primary productivity of about 13×10^3 kg per ha per annum, considerably less than the maximum of 22×10^3 kg per ha. Thus, the current net annual productivity attained when plantations are 20 years old probably represents the maximum for woodland ecosystems with Scots pine as the tree species growing under these conditions. Similar studies in an age sequence of natural birch stands on fen peat (Ovington and Madgwick, 1959b) gave a mean annual increment for the trees alone of just over 6×10^3 kg per ha at 55 years of age and a maximum current net primary productivity of about 8×10^3 kg per ha per annum.

iii. Net Primary Productivity in Different Woodlands. Since annual net primary productivity changes as a woodland matures, it is unwise to make detailed conclusions from comparisons of published data, which on the whole give no indication of the stage of development of a woodland in relation to productivity change. If possible, comparisons of productivity at the time of maximum production would be the most meaningful, but at present the available data are too scanty for this. Volume yield tables give some indication of the differences in maximum productivity that might be anticipated although they only refer to tree boles. Both the current and mean annual volume increments for tree boles, to a top diameter of 3 in and including tree trunks removed as thinnings, are given in standard British yield tables. The volume figures from these tables converted to oven-dry weights for quality class I sites for the following species, *Abies grandis*, *Pseudotsuga taxifolia*, *Pinus nigra*, *Picea abies*, *Pinus sylvestris*, *Larix leptolepis*, *Fagus sylvatica*, *Larix decidua* and *Quercus robur*, give maximum current annual increment figures for the tree boles, expressed as 10^3 kg per ha, as 17, 15, 12, 11, 9, 9, 8, 7 and 6, whilst the maximum mean values are 12, 10, 8, 8, 7, 7, 6, 5 and 4 respectively. The age at which maximum bole production is achieved varies greatly, the peak current annual increment value usually occurs between 15 to 45 years, and the peak mean annual increment value usually occurs between 45 to 80 years. The total annual net primary productivity of the woodlands for all types of plant material would be expected to be much greater than these bole increments, the increases varying for different tree species.

Whittaker (1961) working in the Great Smoky Mountains, U.S.A. gives an annual net primary productivity for *Rhododendron maximum* of 2×10^3 kg per ha and for a mixed shrub community of 10×10^3 kg per ha. Möller *et al.* (1954) have estimated the current annual net primary productivity of beech growing on a second quality site in Denmark to be about 14×10^3 kg per ha at 25 and 46 years of age, but this does not include any allowance for seed production, nor for the ground flora. Ogawa *et al.* (1961) obtained net primary productivities of 19 and 25×10^3 kg per ha per annum for semi-natural temperate evergreen and evergreen gallery forests respectively in Thailand, but point out that these are under-estimates and suggest that the real annual net primary productivity of the tropical forest is probably nearer 40×10^3 kg per ha. They refer to investigations in Japanese forests of *Distylium racemosum* and *Abies sachalinensis* which gave net primary productivities of about 22×10^3 kg per ha per annum and to a thicket of *Arundo donax*, a bamboo-like grass, of the humid tropics with an extraordinarily high annual value of 52×10^3 kg per ha.

iv. Seasonal Pattern of Net Primary Productivity. The rate of net primary production in a woodland ecosystem is not constant throughout the year but exhibits a well-defined seasonal rhythm which may be different for the various plant strata. An extreme example of this is seen in the deciduous woodlands of southern England where bluebell, *Endymion nonscriptus* frequently dominates the ground flora. In March, before the tree buds have opened, the bluebell grows rapidly so that at peak biomass, which occurs towards the end of May, the above ground shoots weigh about 500 kg per ha. The aerial shoots of the bluebell have withered away by early summer when the leaves of the overstorey vegetation are fully developed, so that primary production by bluebell ceases completely until the following spring. During summer and winter some loss of photosynthate through respiration occurs. The initial growth of the bluebell shoots depends on the transference of food reserves from subterranean bulbs and an overall increase in dry weight only takes place from mid-April to the end of May. During this seven-week period the total bluebell biomass trebles in weight, the increase in bulb weight amounting to 60% of the original (Blackman and Rutter, 1947), even though the light intensity at ground level progressively decreases to about 0·3% of that in the open. The intimate phenological synchronization between the over- and understorey vegetation effectively improves the productive capacity of these woodland ecosystems since organic matter production by the bluebell precedes that of the trees.

Marked differences exist in the length of the growing season of trees of different species and in their growth when growing under similar conditions. Kozlowski and Ward (1957a,b) have recorded the pattern of seasonal height growth of conifers and deciduous trees; in some cases height growth was concluded within 10 weeks but in all cases height growth commenced in early spring, and had ended long before the autumn frosts. Kozlowski and Ward do not record whether lammas shoot growth occurred, but in England the production of lammas shoots may account for a large proportion of the height growth and materially extend the growth season (Ovington and MacRae, 1960). The radial growth of tree stems shows a somewhat similar pattern of seasonal growth although radial increment continues for a longer period than height growth (Fritts, 1958; Fraser, 1952). Seasonal growth data do not give a true measure of the seasonal fluctuations in net productivity, thus Kozlowski and Ward suggest that early shoot growth depends on the utilization of photosynthate produced in earlier years and that after height growth has ceased the carbohydrate reserve continues to accumulate. Rutter (1957) studied the annual pattern of dry-weight change for 2- to 5-year-old plants of Scots pine growing in south England and found no general increase in dry weight during the early spring period of rapid elongation of shoots and roots, the greatest weight increments occurring from May to September with a slight increase in dry weight from October to March. Regular seasonal variations in net photosynthesis have also been recorded by field techniques such as those developed by Huber (1950) which are based on measurements of carbon dioxide absorption. Using such a technique Saeki and Nomoto (1958) found that the photosynthetic activity of a deciduous tree *Zelkowa serrata* reached its maximum in April and this was maintained for about 2½ months before decreasing slowly until defoliation occurred in November. Coniferous trees showed a similar seasonal pattern of photosynthesis to *Zelkowa* except that some net photosynthesis occurred in winter although at a much lower rate than during the summer months. There has been some controversy about whether the amount of net photosynthesis of evergreen trees in winter is significant (Bourdeau, 1959). The extent of winter curtailment of net photosynthesis depends upon the severity of the cold weather and the cold-hardiness of the trees (Parker, 1961) so that when climatic conditions are favourable, evergreen trees give small but positive winter net photosynthesis rates. Pearson and Lawrence (1958) have shown that large amounts of chlorophyll are contained in the bark of the quaking aspen *Populus tremuloides* and suggest that in early spring, before leaf emergence, photosynthesis may occur in the tree bark of deciduous hardwoods.

2. *Gross Primary Production*

Since part of the photosynthate produced by woodland plants is used in their respiration, the gross primary productivity is in excess of the net values. The rate of respiration varies greatly for different plant parts, being high for leaves and relatively low for woody material such as tree trunks, and also shows marked seasonal changes (Johansson, 1933). Möller *et al.* (1954) have calculated the annual loss of photosynthate by respiration for beech stands in Denmark, expressed as 10^3 kg per ha, to be about 6 at 8 years of age and increasing to 10 at 85 years, equal to 85% and 89% of the net primary productivity. Ogawa *et al.* (1961) using data for Japanese forests give respiration losses of 243% and 124% of the net primary productivity for *Distylium racemosum* and *Abies sachalinensis* forest, whilst Tranquillini (1959a,b) found that for 5- to 8-year-old trees of *Pinus cembra*, the loss of photosynthate by respiration was about 62% of the annual dry matter increment.

3. *Woodland Ecosystems as Productive Units*

i. Woodlands Compared with other Terrestrial Ecosystems. Natural and artificial ecosystems containing green plants all have the ability to synthesize and accumulate organic matter and it is of interest to contrast woodlands with other ecosystems for the efficiency with which site conditions are utilized for the annual production of organic matter. The main difficulty is to decide on the most appropriate basis on which to assess woodland productivity. At present there are few data available of gross productivity, so that net primary productivity seems the most suitable basis for comparison. A distinction can be made between primary biological productivity and the production of plant matter of economic value (Ničiporovič, 1960). Of these the economic yield is the least satisfactory value for fundamental comparisons, since the intensity of harvesting and type of plant material cropped are liable to change with changing social conditions. Valid reasons can be given for presenting net primary biological productivity as either the greatest values for the mean or current annual production and doubt may be expressed about the use of both values. It can be argued that the period of maximum biological production is a relatively short phase in a forest rotation and cannot be maintained on a long-term basis so that the true measure of forest productivity is the seasonal turnover of organic matter at the equilibrium stage when no significant change in biomass is taking place from year to year. An alternative basis for comparison is the rate of accretion of organic matter into woodland ecosystems, possibly with some addition for the wood harvested. Tentative figures for some of these different measures are given in Table V; it should be emphasized that these tend to approximate to the maximum values for temperate regions since they are based on the available figures for high production woodlands growing on good sites.

There are few determinations of the net primary productivity of other types of uncultivated terrestrial ecosystems. Billings and Bliss (1959) have examined various characteristic alpine ecosystems and record the following rates of net primary production of dry matter exclusive of roots, expressed as 10^3 kg per ha: alpine snowbank communities 0·3 to 1·3, alpine meadows 0·2 to 1·1, an herbaceous community in North Alaska 0·6, heath *Juncus* field 1·3, heath *Juncus*

Table V

Approximate Maximum Primary Production of Organic Matter by High-producing Woodlands in West Europe

	Oven-dry weight 10^3 kg per ha
Current gross production	40
Current net production (Biological)	22
Current net production (Economic-tree boles)	17
Current accretion of organic matter	10
Mean gross production	20
Mean net production (Biological)	15
Mean net production (Economic-tree boles)	12
Mean accretion of organic matter	7

meadow 2·4 and *Carex* meadow 3·4. Specht *et al.* (1958) in South Australia have made intensive studies of the heath vegetation succession following fire and found that the annual biomass increment of plant material, both roots and tops, was $3{\cdot}2 \times 10^3$ kg per ha from 9 to 25 years — primary production must exceed this. Odum (1960) has determined net production in old abandoned fields in South Carolina, U.S.A.; in the first year after abandonment the invading flora produced 5×10^3 kg of dry matter per ha, but in the following 6 years annual production decreased to about 3×10^3. Bray *et al.* (1959) give a range of yearly organic-matter production for some terrestrial communities in Minnesota of from 5 for *Setaria glauca* to 46 for *Typha* although they recognize that part of this organic matter was derived from previous growing seasons. A coral-reef community on Eniwetok Atoll had the large annual productivity of 81×10^3 kg per ha (Odum and Odum, 1955) but this may be regarded as semi-marine; fairly high production rates have been recorded for aquatic communities, e.g. Odum (1957) gives an annual dry-weight production at Silver Springs in Florida of 65×10^3 kg per ha.

Numerous data of the productivity of agricultural crops are available but need cautious interpretation since they frequently refer only to the marketable plant parts. Production by weeds is frequently ignored. Transeau (1926) for example has made an extremely detailed study of production by maize in Wisconsin, U.S.A., preparing energy balance-sheets for a field of maize with an annual net primary productivity of 15×10^3 kg per ha but does not mention the productivity of the weed flora. In the neighbouring state of Minnesota, maize produced 9×10^3 and the associated weeds 1×10^3 kg of dry matter per ha per annum on rather poor farmland. On the better soils of the corn belt much greater productivity is generally attained. Armstrong (1960) reports an annual cut, at probably the period of maximum biomass, of 6×10^3 kg per ha from rye grass at Hurley in England but gives no figures for root production. Russell (1956) has given the following dry weights, expressed as 10^3 kg per ha, for crop production commonly obtained on farms in England, beans 3·9, meadow hay 3·2, red clover hay 4·2, oats 4·5, wheat 4·7, turnips 5·2 and mangolds 8·5. The maximum yields on good soils with adequate fertilizers and water would be expected to exceed these values. Since Russell does not give any values for weed production and the root mass is not included in these figures, they underestimate the total net productivity, probably by as much as a half.

It is evident from the available records that, compared with other terrestrial communities, woodlands are fairly efficient long-term producers of organic matter. Estimates by de Wit (1959) based on known rates of photosynthesis and average measurements of incident solar radiation show that, for the 6 months, April to September, the maximum possible net primary production for a closed green crop in the Netherlands is about 36×10^3 kg per ha, so that West European forest production is considerably less than the potential production indicated by incoming solar radiation. Restrictions on production may be imposed by factors such as water and nutrient shortage. The high level of production of woodlands is achieved with remarkably little input of human energy since forest soils are not normally ploughed or fertilized regularly and the growing crop receives comparatively little attention. In addition, the stability of the forest ecosystem is such that harvesting of large amounts of organic matter as wood is possible without greatly affecting the overall woodland productivity, at least over several generations of trees. Ničiporovič and Strogonova (1957) have suggested that good economic yields are dependent upon large biological production and this is the case for woodlands. The chemical composition of the organic matter produced by different ecosystems varies greatly, the forest production being essentially bulk carbohydrate rather than protein. It seems pertinent to consider what the factors are that limit primary production in woodland ecosystems.

ii. Factors Affecting Primary Production in Woodlands. Organic matter production by woodlands may be considered from three interrelated viewpoints, (*a*) the environmental limits to productivity, (*b*) the features of woodland ecosystems enabling them to attain a relatively high level of efficiency in exploiting the environment, and (*c*) the possibilities of improving productivity by sylvicultural operations.

Several environmental factors are known to limit woodland productivity and there is an extensive literature describing the circumstances in which a particular factor becomes limiting. Kramer (1958) has recently reported on the present status of research relating to photosynthesis by trees with particular reference to the influence of the more important environmental factors namely, light, temperature, moisture, mineral nutrition, atmospheric carbon dioxide, defoliating insects, pathogenic fungi and toxic gases released by industry. Since photosynthesis is the plant metabolic process most directly concerned in the production of organic matter, the same factors, with the addition of wind essentially control the upper limit of productivity, i.e. the site capacity.

Within the limits set by the environment, the rate of organic matter production by woodland and other terrestrial ecosystems varies enormously but attention has been drawn to the relative efficiency of some woodlands. Working with agricultural crops Watson (1958) concluded that with an adequate supply of moisture and minerals, the rate of organic matter production depends upon the efficiency and more particularly the size of the photosynthetic system. So far as size is concerned, woodlands normally have a relatively large mass of green photosynthetic tissue. This is primarily tree leaves, the maximum weight of tree leaves recorded in Table I being $19{\cdot}8 \times 10^3$ kg per ha, but photosynthesis also occurs in the tree twigs and fruiting bodies and in the understorey vegetation. The leaves of the woodland flora have a large surface area, Tadaki and Shidei (1960) give a range of tree leaf areas from 2·2 to 7·9 ha per ha of land surface for stands of deciduous tree species. Much of the incident light is intercepted and there is a progressive decrease in light intensity through the successive plant layers until the light at ground level may only amount to about 1% of that falling on the ecosystem. Passing downwards from the upper plant surface of the ecosystem there is a gradation in photosynthetic efficiency from sun to shade-leaves and to the shade-tolerant plants of the ground flora which presumably benefit from the increased carbon dioxide concentration at ground level. Large scale absorption of carbon dioxide from the atmosphere is necessary for high productivity and would be facilitated by having the large area of chlorophyll-bearing tissue dispersed through a considerable depth of air. Evidently, the relatively high rate of organic matter production attained by woodlands can be partly attributed to the large and well-distributed mass of photosynthetic tissue. Ultimately the upward extension of the photosynthetic zone might be expected to lower the efficiency of organic matter production since with increasing height more energy will be expended in translocation and in transporting water from the roots to the leaves. It is significant that the greatest annual production of organic matter is achieved at about the pole stage when the trees are about 7 m tall and the weight of tree leaves present per unit area of land is at a maximum and the living canopy is just being raised clear of the ground (Ovington, 1957a).

Any evaluation of the potential capacity of an ecosystem to manufacture organic matter also needs to take into account the annual duration of photosynthesis. This can be illustrated by comparing forest and agricultural crops. Agricultural crops such as spring cereals may attain high daily rates of net photosynthesis but the annual production of plant matter may still be low compared to that of woodland ecosystems because of the short annual life history of the cereals. Frequently agricultural crops are only present for about five months of the year and only during part of this time do they make their fullest use of the site, since time is required for the leaf and root systems to develop from the planted seed and, photosynthesis of the crop is greatly reduced during the ripening period. In contrast, evergreen woodlands have a large weight of leaves and roots present throughout the year so that positive values for net photosynthesis may be recorded even in the winter period. In Britain, deciduous woodlands tend to produce less organic matter per annum than evergreen woodlands and this difference can be attributed in part to differences in duration of the photosynthetic period. Nevertheless, deciduous woodlands appear to have some advantages over agricultural crops in their organic matter production, since the root mass persists throughout the winter, chlorophyll is present in the branches during the over-wintering period and leaves can be produced fairly rapidly in spring from the mass of dormant buds in the tree crowns. The great diversity of plant form and phenology in woodland ecosystems compared with the uniformity of an agricultural field is another important factor affecting the production of organic matter. The complementary role of the ground flora in extending the period of organic matter production in deciduous woodlands has been mentioned previously.

Large amounts of organic matter cannot be manufactured without adequate supplies of water and plant nutrients and it is significant that the root systems of woodlands normally tend to penetrate a large soil volume from which they can draw water and nutrients. There is some evidence that root systems of individual trees do not function independently, since root grafting between forest trees is frequently observed and it appears that many grafts permit the translocation of substances and movement of water between individual trees (Bormann and Graham, 1959). It has been suggested that when the aerial parts of a tree are killed or harvested its root system may persist and contribute to the growth of the surviving trees. This helps to explain the rapidity with which woodlands recover from thinning. It seems clear that woodland ecosystems have certain attributes, partly a result of the genetical constitution of the individual plants but also of the characteristic multistoreyed structure of the ecosystem giving them a capacity to manufacture a relatively large amount of organic matter annually. The dense stands of high producing conifers that have been established in upland Britain within the last two decades constitute a practical demonstration of the relatively great potential of woodland ecosystems to produce plant matter even on difficult sites.

The evidence from numerous field trials indicates that there is considerable opportunity for sylviculturalists to increase the production of plant matter by woodland ecosystems, although it is important to distinguish between short- and long-term effects as well as differences in the production of tree boles and plant matter in general. The initial phase of low productivity before the tree crowns meet can be greatly shortened by planting or favouring more vigorous tree species and by boosting initial tree growth with establishment techniques, such as weeding, soil cultivation and manuring.

Once true woodland conditions are established, woodland ecosystems are capable of making fairly full use of the site since the peak annual biological productivity of a high producing forest type presumably corresponds closely to the productive capacity of the site. Evidence to support this is found in the interrelationships between the over- and understorey plants, the extinction of the ground flora at the pole stage indicating that the production limits of the site have been attained by the trees alone. At this stage the scope for increasing biological productivity is more restricted, the most promising approach being to try and improve site conditions.

Some environmental factors can be changed easily. When lack of mineral nutrients limits productivity, various improvement techniques may be adopted, e.g. the application of fertilizers, the breaking of impenetrable soil layers by ploughing in order to make a larger volume of soil available and the use of appropriate mixtures of complementary tree species (Pogrebnyak, 1960). Similarly soil moisture can be regulated by irrigation or drainage and pest control attempted by biological or chemical means.

Other environmental factors such as the carbon dioxide concentration of the air, temperature and light cannot be changed readily over large areas, and when these factors are limiting it seems that biological productivity can only be increased greatly by breeding plant varieties having a greater photosynthetic efficiency. Nevertheless, Dadykin (1960) has suggested that some improvement in production can be obtained by running the tree rows from north to south and using certain combinations of tree species with canopy characteristics giving the maximum absorption of the incident solar energy.

Whilst it is generally accepted that sylvicultural operations such as thinning and pruning result in an improvement of timber quality, their effect on productivity has been the subject of some controversy. After a detailed examination of some of the experimental data Möller (1954, 1960) concluded that within wide limits of thinning intensity, the average increment of the trees over long periods is not affected by thinning and that the long-established maxim that removing the lower third of the live crown of trees does not reduce the increment is incorrect. However, Möller is primarily concerned with trees, particularly tree boles, and not with whole ecosystems. In a thinning experiment with Scots pine in Britain, at 29 years of age no ground flora was present in the lightly thinned plot but in the heaviest thinned plot the ground flora had an oven-dry weight of $0{\cdot}6 \times 10^3$ kg per ha, so that if the productivity of the trees is constant the productivity of the ecosystem must have been changed by thinning. Haberland and Wilde (1961) have shown that heavy thinning in dense red pine plantations in Wisconsin induces unfavourable soil conditions so that in the long run any immediate effect on tree growth is offset by deterioration of the site. Comprehensive ecosystem studies are needed to elucidate the basic biological background of thinning and pruning.

Great advances in tree breeding have been made in the last quarter of a century (Larsen, 1956) and undoubtedly plant breeders have an important role to play in improving productivity. Biological productivity can be increased by breeding new varieties or by careful selection within existing woodlands to obtain trees easy to establish, capable of high rates of photosynthesis, and resistant to disease, drought and cold. There is also considerable scope to grow trees in which a greater proportion of the biological production is diverted to salable tree trunks, i.e. to economic production.

D. ORGANIC MATTER TURNOVER

Only part of the organic matter produced by woodland plants is retained within the ecosystem, the rest being lost in various ways. Man, for example, removes all kinds of forest produce and the intensity of his harvesting varies enormously.

Organic matter is also removed from woodland ecosystems by natural means, e.g. by plant roots growing beyond the recognized ecosystem limits, by animal movement out of the ecosystem and by wind dispersal of seeds, pollen and leaves. Little attention has been paid to the natural flow of organic matter into and out of woodland ecosystems and frequently it is assumed incorrectly that these are in balance and insignificant in quantity. Sjörs (1954) made an intensive study of some Swedish park meadows having about 25% of the area wooded and found that the yearly fall of tree leaves in the woodland varied from 1·5 to 2·5 × 10^3 kg per ha. Depending on the size of the opening, the treeless areas received 0·3 to 0·8 × 10^3 kg of leaves per ha per annum from the woodland. In extreme cases, such as exposed savanna or narrow shelter belt type woodlands, soon after leaf fall the woodland floor may be cleared completely of tree leaves by the wind, if the understorey vegetation is ineffective in trapping and retaining the annual leaf fall.

Normally, the greatest loss of organic matter results from decomposition within the ecosystem and this far exceeds that due to harvesting (Fig. 4). Comminution of the primary woodland production begins on the living plants and continues in the soil after litter fall. Litter fall takes place throughout the year in both coniferous (Owen, 1954) and deciduous (Miller and Hurst, 1957) woodlands. The breakdown of organic material involves various agencies including the fauna, microflora and environmental factors such as frost and wind. Woodland ecologists have paid particular attention to the complementary activities

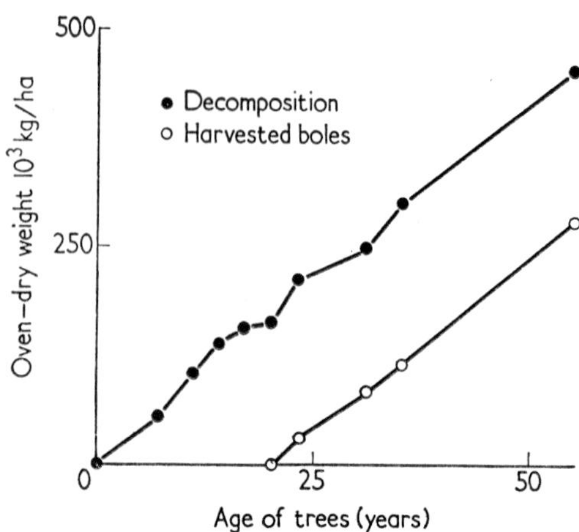

FIG. 4. Loss of organic matter by decomposition and harvesting of tree boles in plantations of *Pinus sylvestris*.

of the soil micro-fauna and -flora in litter decomposition. Soil animals are known to affect litter breakdown in numerous ways: by physically breaking up litter, by chemical modification of the plant material as it passes through their intestines, by mixing organic matter with the mineral soil and by feeding on the microflora.

Tamm and Östlund (1960), using radiocarbon dating, in a region where forest has been dominant for 9000 years, found that the organic matter in the B soil horizon of a 95-year-old spruce forest had been present for up to 370 (± 100) years. The turnover of most organic matter occurs more quickly than this and recently, fairly rapid rates of breakdown have been recorded by tracing the progressive loss in weight of tree leaves placed on the ground in nylon net bags (Bocock and Gilbert, 1957; Bocock *et al.*, 1960; Shanks and Olson, 1961). This technique has shown that the rate of leaf breakdown depends on the tree species, the chemical composition of the leaves, soil conditions, moisture and temperature.

Tree leaves only constitute part of the litter fall, which consists of a complex and changing mixture of plant and animal material of varied structure and chemical composition (Fig. 5). Typical components are dead animals, animal frass, the remains of the understorey plants, cones, bark, twigs, etc.; all of these decompose at very different rates from the tree leaves (Ovington, 1959a). The influence of the diverse, mixed nature of the litter fall on the overall rate of decomposition has not been fully appreciated, yet the physical, chemical and biological nature of the litter mass depends on the relative proportions of the different components. In nature no single litter component can be considered in isolation since, between the various components, there are many interactions which affect the overall rate of decay. The composition of the litter horizons overlying the mineral soil, changes considerably, both as a result of variations in the litter fall at different stages of woodland development, and because of the differential rates of decomposition of the types of organic matter present. For example, pine cones decompose more slowly than pine needles so that the proportion of cones in the litter on the soil becomes greater than would be expected from changes in cone content of the litter fall alone (Fig. 5).

Measurements of the rate of breakdown of litter over short periods do not give a true evaluation of comparative differences between woodland ecosystems in the mass annual turnover of organic matter through the litter. Lull (1959) has compared humus depths in virgin and managed forest stands in the north-eastern Appalachian Mountains and found that the depth of humus in old managed stands was only slightly

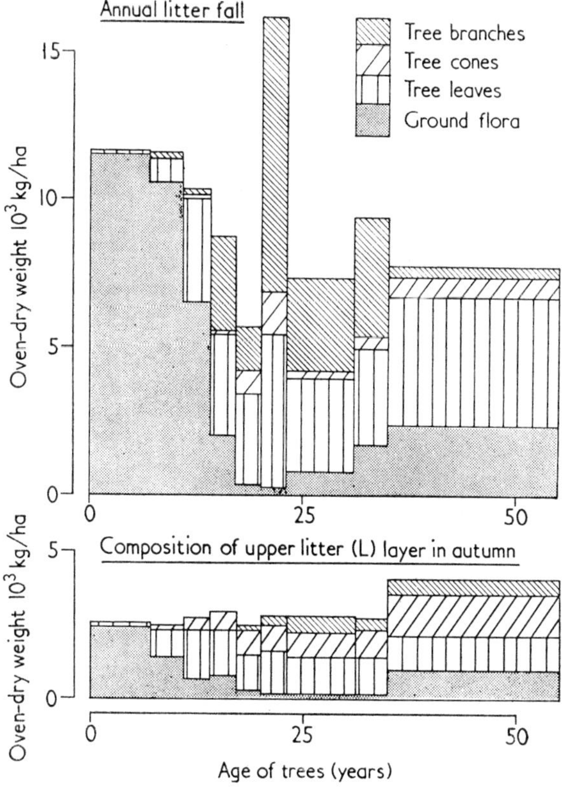

Fig. 5. Litter fall and composition of surface litter in plantations of *Pinus sylvestris*.

less than in corresponding virgin stands. He suggests that there is a maximum humus depth at which equilibrium is attained between annual litter fall and decomposition. It is noteworthy that although dead organic matter accumulates rapidly over the ground in young coniferous plantations, ultimately a stage is reached at which no further litter accretion takes place. Once the weight of the litter layers becomes constant from year to year, annual litter fall and decomposition must be equal so that the annual litter fall provides a measure of organic matter breakdown at the surface of the mineral soil.

From the limited viewpoint of the mass annual turnover of the primary production of aerial shoots, once the equilibrium stage between litter fall and decomposition is reached, there is little significant difference between woodlands having very different amounts of litter accumulated on the ground so long as the amount of litter fall is equal. Coniferous forests in Britain frequently develop a thick layer of organic matter over the mineral soil, about three to five times the weight of the annual litter fall, and we can assume that on the whole, the annual litter fall takes from 3 to 5 years to decompose. In deciduous woodlands at similar sites, there is frequently little carry over of litter from year to year so that annually complete decomposition of the litter fall is achieved. In fact decomposition is often so rapid that it is virtually completed within a 6- to 9-month period. Nevertheless, since litter fall is often much greater in softwood than hardwood forests, the annual rate of organic matter turnover through the coniferous litter may far exceed that of the deciduous hardwood forest.

Studies of organic matter turnover have been largely concerned with the above-ground parts but the turnover of roots may be equally important. Harley (1959) has published a series of photographs of beech roots at different times of the year and points out the seasonal loss of roots which he attributes to drought or animal action. Unfortunately few quantitative data of root decomposition are available but Remezov (1959) reports that in a 50-year-old oak stand in U.S.S.R. the mortality of roots less than 3 mm in diameter amounted to about $0{\cdot}3 \times 10^3$ kg per ha from June to August. Orlov (1955) has reported that the turnover of roots in a 25-year-old conifer stand was about 50% of that of the needles and in a 50-year-old stand about 20%.

In the older plantations of the age series of Scots pine, the annual turnover of plant material through the litter amounts to about 8×10^3 kg per ha and, if allowance is made for the breakdown of small roots, the annual turnover of organic matter would be over 9×10^3 kg per ha. Other more productive woodlands would be expected to have a greater annual decomposition of organic matter; Ogawa *et al.* (1961) for example, give a range of values increasing to 25×10^3 kg per ha for the evergreen gallery forest.

IV. Energy Dynamics

The woodland plants capable of photosynthesis transform part of the solar energy reaching woodland ecosystems to chemical energy, which is stored in the plant organs. Some of the energy may be dissipated later through the metabolism of the producer plants, the rest acts as a source of energy for the other living organisms of the ecosystem. Whilst the overall pattern of the energy system broadly parallels that of the organic system, the two do not correspond exactly since the energy content of organic material is variable. Golley (1961) has stressed the relative constancy of the calorific value of animal tissue at about 5 000 cal per g dry weight compared with that of plant material which shows considerable variation with an average of about 4 500 cal per g dry weight.

As organic matter is broken down by animals and heterotrophic plants, the energy it contains is either released or transferred along the food chain in the consuming organisms, each organism depending upon the preceding living organisms for a supply of energy. Because of the respiration of the consumer organisms, the amount of energy transferred along the food chain becomes progressively smaller as comminution proceeds. When organic matter accumulates within woodland ecosystems, such as young plantations, the annual fixation of energy exceeds energy release so that there is a carry over of energy from year to year and eventually, a large store of energy may be built up in the ecosystem. The detailed pattern of energy fixation, storage, flow and release in woodlands is extremely complicated and is incompletely understood but it is possible to prepare energy budgets for the plant material showing in a general way the magnitude of the major features of the energy system in woodlands (Fig. 6).

A. Energy Fixation

Whilst certain micro-organisms are capable of chemosynthesis and may contribute energy to woodland ecosystems, chlorophyll-bearing plants are virtually the sole agents of energy fixation and the primary organic matter produced by the trees, shrubs and ground flora is the principal source of energy on which the rest of the living organisms ultimately depend. Consequently, the photosynthetic efficiency of the green woodland plants places an upper limit to the potential metabolism of the entire ecosystem. It has been suggested that woodland ecosystems are efficient producers of organic matter compared with other types of terrestrial ecosystems and attention has been drawn to the low light intensity at ground level in dense canopied woodlands, so that comparatively high levels of energy fixation by woodland ecosystems may be anticipated.

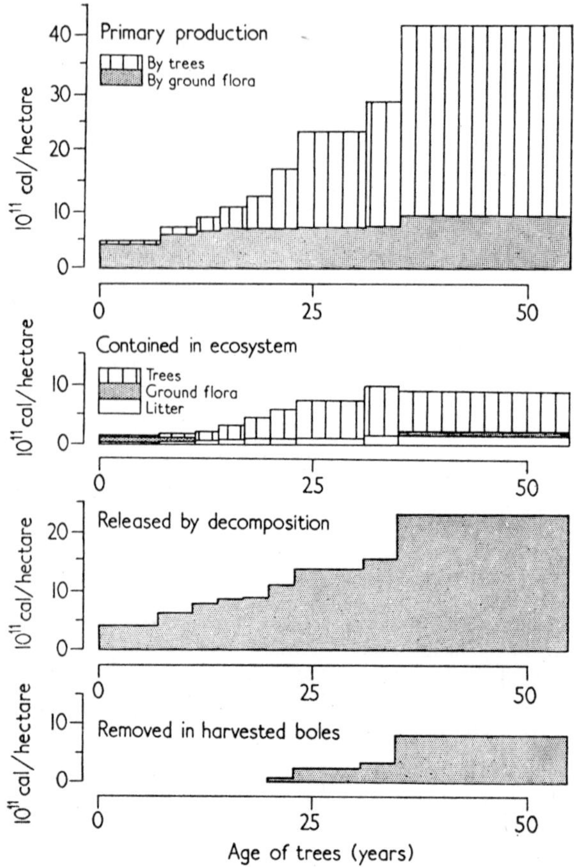

FIG. 6. Energy budget for plantations of *Pinus sylvestris* from time of planting.

The magnitude of energy flow through the series of Scots pine woodlands referred to earlier (Ovington, 1961) has been determined by calorimetry. During the 55-year period following tree planting, the primary net production per ha contained just over 422×10^{10} cal, equivalent to a mean annual transformation of solar energy of about 8×10^{10} cal, of which just over three-quarters can be attributed to the trees (Fig. 6). At the peak period of organic matter production when the trees are essentially the sole primary producers, the annual fixation of energy in the net primary production is 10×10^{10} cal per ha. Since the average annual income of solar radiation to the pine plantations is about 770×10^{10} cal per ha of land surface, about 1·3% of the incident energy is stored within the primary products of photosynthesis, and this is equivalent to about 2·5% of the annual incident radiation occurring within the radiation wavelengths known to be utilized in photosynthesis (4 000 to 7 000 Å). Wassink (1959) has concluded that, when incident radiation values are limited to the wavelengths available for photosynthesis, the overall annual photosynthetic efficiency for field crops in Holland lies between 1 and 2%, although over periods of a few weeks efficiencies of 7 to 9% were attained. Scots pine is not the most productive of tree species and the site conditions are not of the best, so that other woodlands elsewhere in Britain attain higher rates of production and of energy transformation, with an annual efficiency of energy fixation probably just over 3%. Tropical forests at least equal temperate forests in photosynthetic efficiency. The estimate of Ogawa *et al.* (1961) of a net annual production of dry matter of 40×10^3 kg per ha for tropical forest gives an annual fixation of about 18×10^{10} cal per ha in a region where the annual incident radiation within the range for photosynthesis is about 600×10^{10} cal per ha, thus photosynthetic efficiency is around 3%. This compares well with 1·9% for sugar-cane in Java given by Hellmers and Bonner (1959). Monteith (1959) has suggested that forests may absorb more radiation than agricultural crops because of their darker colour and the trapping of radiation between individual tree crowns and conifer needles so that reflection of radiation is at a minimum. Thus a pine forest in northern Scotland may absorb more radiation than a pasture in southern England, even though the incoming radiation is less, because of the lower amount of reflection from the pine canopy.

Of the sunlight penetrating through the earth's atmosphere, only a small portion is converted to chemical energy by photosynthesis. This is not surprising for solar energy is dissipated in various other ways, and the photosynthetic process is not 100% efficient. Energy capture by the faster-growing woodland ecosystems at their most productive period may therefore be approaching the maximum possible under natural conditions and so provides a means of determining maximum site potential, a value of considerable interest to ecologists and of some application to land use projects aimed at improving productivity.

B. ENERGY ACCUMULATION AND RELEASE

The storage of energy in the whole woodland ecosystem is considerable. When forest fires break out, this energy is released rapidly as heat and very high temperatures result, despite the rapid dissipation of heat upwards. Davis (1959) states that in hot forest fires, the above-ground temperatures in the organic mass may exceed 800° C. Temperatures of 200° C and higher are common at the ground surface and, with much organic matter present, burning may continue for over two hours so that soil conditions may be changed.

In a variety of British woodlands of from 18 to 47 years of age, the amount of energy in the plant organic matter (exclusive of roots) ranged from 22 to 156×10^{10} cal per ha depending upon the tree species and the development of the woodland (Ovington and Heitkamp, 1960). In neighbouring treeless areas the energy content of the vegetation is much smaller varying from 2 to 6×10^{10} cal per ha. Most of the energy in woodland ecosystems is contained within the trees. The maximum mean annual rate of energy accumulation recorded for these woodlands was 7×10^{10} cal per ha for a young plantation of *Picea omorika*, equivalent to about 1% of the incoming solar radiation, and more than double the accretion of energy in Scots pine plantations of similar age. In the older 47-year-old plantations, harvesting of the tree boles has removed up to 100×10^{10} cal per ha, i.e. just about one-eighth of the incident solar radiation of a single year. Clear felling and complete harvesting of the boles would result in a total energy removal of about 190×10^{10} cal per ha. Unfortunately no data are available of the primary productivity of these woodlands but in Table VI an energy budget is given for the relatively slow-growing Scots pine plantations during their most productive period. The estimate of the respiration of the trees and ground flora is calculated from Möller's figures for beech and gives an energy release by respiration of the producer plants of about 160×10^{10} cal per ha for the 18-year period.

Over the 18-year period covered in Table VI, the annual breakdown of the net primary production results in an average annual liberation of energy of about 4×10^{10} cal per ha by the animals and heterotrophic

TABLE VI

Energy Flow following Photosynthesis of Scots Pine Plantations for the 18-year Period from 17 to 35 Years after Planting

	Energy content 10^{10} cal per ha	Percentage utilization of total incident solar radiation
In gross primary plant production	340	2·46
Released by respiration of producer plants	160	1·16
In net primary plant production	180	1·30
Accumulation in living trees	61	0·44
Accumulation in ground flora	<1	$<0·01$
Accumulation in litter layers	6	0·04
Removed in boles of harvested trees	31	0·23
Left in roots of harvested trees	13	0·09
Released by litter decomposition	68	0·49

plants. This is over twice the annual energy released *per capita* in Britain from all kinds of industrial fuel. Tentative estimates of the energy content of the woodland fauna and heterotrophic plants, from the fauna weights in Table III and microflora numbers, indicate that their maximum possible energy content is in the order of $0{\cdot}2 \times 10^{10}$ cal per ha, about one-twentieth of the annual energy release. Since the biomass figures for the fauna and heterotrophic plants are based on the maximum values, the total is probably too high and the true ratio of energy content to annual flow is probably nearer one two-hundredth (Macfadyen, 1961).

The preparation of energy-balance tables showing the detailed accounting of energy flows of this magnitude by the consumer organisms in different types of woodland ecosystems is not possible at present but represents one of the most important and challenging problems to woodland ecologists. A recent investigation by Golley (1960) illustrates the need to combine closely both botanical and zoological studies in order to understand the functioning of terrestrial ecosystems in relation to energy dynamics.

V. Water Circulation

On a world scale, the hydrologic cycle involves the interchange of water between the oceans and the land *via* the atmosphere. Sutcliffe (1956) has pointed out that the amount of precipitable water contained in the world's atmosphere only amounts to 2·5 cm of rain, i.e. 25×10^4 kg of water per ha of the earth's surface, so that a huge and rapid transfer of water takes place continually. Normally, precipitation, as rain or snow, constitutes the main source of water for forest ecosystems, although in some circumstances this may be supplemented by the upward movement of subterranean water and by the lateral flow of soil water from neighbouring ecosystems. Woodlands rarely occur in regions where the annual input of water is less than 300×10^4 kg per ha. Whilst it is unlikely that forests materially modify the mass global circulation of water, it has been implied that they may increase precipitation locally by inducing condensation through a reduction of air temperature and an increase of humidity, both a result of transpiration, and by collecting water droplets from low-lying cloud and fog (Oberlander, 1956; Bleasdale, 1957). Apart from areas where cloud- and fog-drip take place, it seems doubtful that the presence of woodland significantly changes the total precipitation, but once precipitation occurs, the nature of the forest cover greatly affects the passage of water through the ecosystem.

The pattern of water circulation in woodland ecosystems does not follow the organic system so closely as does energy flow or mineral circulation. The two most significant features of the woodland water cycle are, (i) the relatively rapid passage of large amounts of water through the ecosystem and, (ii) the small amount of water retained within the ecosystem, particularly in the organic matter. This is shown by Table VII where the distribution of water in late summer is given for three softwood and three hardwood plantations in Britain, all growing on very similar sites and in close proximity to one another. After 47 years of afforestation, the water present in the forest ecosystems only averages about 15% of the annual precipitation and less than 0·1% of the annual precipitation is contained within the organic matter accreted each year. Seasonal changes in the water content of the woodland ecosystems occur, e.g. the water content of the mineral soil increases during the wetter, winter months, but if this trebled, the total water in the ecosystem would still be small compared with precipitation. Another interesting feature shown by Table VII is the greater weight of water in the mineral soils of the hardwood plantations. Conversely, the organic matter of the softwood plantations contains more water than that of the less productive hardwood stands. Differences in the water content of the organic matter of softwood and hardwood plots are insufficient to account for the larger differences in mineral soil moisture, which must be attributed mainly to a greater annual loss of water from those ecosystems having coniferous trees present. Water is known to be removed from woodland ecosystems in several ways and considerable scientific effort has been devoted to the measurement of different forms of water loss, mainly from the point of view of slowing the speed of run-off to reduce soil erosion and of increasing the amount of water directly available for use by man, i.e. the water yield.

A. WATER LOSS FROM WOODLAND ECOSYSTEMS

Since the amount of water likely to be removed from woodland ecosystems by cropping is relatively insignificant, water loss occurs mainly through evaporation, transpiration and run-off.

1. Evaporation

Much of the precipitation is intercepted by the woodland vegetation: of this, part is evaporated directly back into the atmosphere, some is absorbed by the vegetation and the rest reaches the ground as water flowing down the plant stems or dripping from the foliage.

Numerous data are available giving the proportion of the annual precipitation failing to reach ground level because of interception and the following are typical; Loblolly pine 14% (Hoover, 1953), 10% to 55% for a range of British plantations (Ovington, 1954b), 35% in mixed Uganda tropical forest (Hopkins, 1960) and 19% to 33% for Red pine

TABLE VII

Water Content (10^3 kg per ha) *of Neighbouring Woodland Plantations*

Tree species	*Pseudotsuga taxifolia*	*Picea abies*	*Pinus sylvestris*	*Fagus sylvatica*	*Quercus robur*	*Castanea sativa*
Age of trees in years	47	47	47	39	47	47
Tree canopy	48	30	34	33	29	12
Tree trunks	151	124	103	68	82	119
Understorey vegetation	8	0	18	<1	3	2
Plant material overlying mineral soil	4	10	9	7	3	4
Total in plant material above mineral soil	211	164	164	108	117	137
Mineral soil (sampled to 70 cm)	796	1 116	1 076	1 242	1 478	1 596
Water content of ecosystem (exclusive of roots and fauna)	1 007	1 280	1 240	1 350	1 595	1 733
Water removed in harvested boles	154	242	170	44	43	65
Water content of above ground plant material as a percentage of annual precipitation (9 220 10^3 kg per ha)	2	2	2	1	1	1
Water content of ecosystem as a percentage of annual precipitation (9 220 10^3 kg per ha)	11	14	13	15	17	19

and beech stands (Voigt, 1960). The magnitude of the interception-loss depends upon both weather conditions and the luxuriance and form of the vegetation cover. Evaporation is more rapid in dry, hot, windy weather. The percentage loss of water by interception is greater in light than heavy showers so that less water is lost by interception if the precipitation consists of heavy downpours. Canopies of coniferous trees with their many needle-like leaves trap more water than those of deciduous hardwoods as the latter have large, flat leaves from which water runs off easily. Interception also results in a more patchy distribution of water over the forest floor (Ovington, 1954b) and this is probably of great importance to the water economy of the trees since water penetration into the soil is improved and the water flowing down the tree-stems tends to follow the roots spreading out from the base of the trunk (Voigt, 1960).

While much of the water failing to reach ground level is evaporated directly back into the atmosphere, there is some evidence that significant amounts of water are absorbed by the foliage so that interception does not represent a complete water loss from the ecosystem. Härtel and Rudolph (1953) found that when needles of *Abies alba* are moistened, between 3 to 10 mg of water are absorbed per g of leaf weight per hour and in young leaves the rate of absorption rises to 30 mg. Stålfelt (1944) calculated that under the most favourable conditions, the absorption of water by spruce leaves is equivalent to 33% of the total transpiration but Rutter (1959) found cuticular absorption of water by Scots pine needles to be negligible.

Once precipitation reaches ground level, further loss of water may occur by evaporation, the amount depending upon the closeness of the forest cover and the rapidity of infiltration into the soil.

2. *Transpiration*

Transpiration of the woodland plants is controlled by two main sets of factors, (i) environmental, e.g. wind, atmospheric humidity and temperature, and (ii) internal plant controls such as leaf structure and root distribution. Ladefoged (1956) found that during June the daily transpiration of a first quality beech stand with trees 25 to 30 years old, could be closely related to weather conditions, thus in wet weather, under overcast skies and in bright sunshine, 0, 20 and 40×10^3 kg of water respectively were transpired per ha. Woodlands of different tree species growing under identical conditions transpire at different rates. Polster (1950) gives comparative figures for the transpiration rates of trees of various species based on measurements of water-loss from twigs of leaves cut from the tree crowns, if the transpiration rate of spruce is given a value of 100 the comparative figures for birch, oak, beech,

Douglas fir, pine and larch become 740, 460, 372, 94, 139 and 212. By combining these data with foliage biomass figures such as those of Burger, Polster obtained the following mean daily transpiration rates for the trees expressed as 10^3 kg of water per ha: birch 47, beech 38, larch 47, Douglas fir 53, spruce 43 and pine 24. The results of Ivanov *et al.* (1951) indicate that the amount of water transpired by trees varies as a woodland matures, in 10, 33, 65, 80 and 150-year-old pine stands, the daily transpiration from May to September (the transpiration period according to Ivanov *et al.*) amounts to 17, 23, 18, 17 and 13 $\times$ 10^3 kg of water per ha compared with 22, 22 and 19 $\times$ 10^3 for birch stands 25, 60 and 70 years old. Apparently maximum transpiration is achieved about the peak period of organic matter production.

Little attention has been paid to the water loss attributable to interception and transpiration by the woodland understorey vegetation but in some circumstances this may be considerable. The removal of the dense shrub understorey of rhododendron and laurel from a wooded catchment in North Carolina increased the annual water yield by 50 $\times$ 10^4 kg per ha (Johnson and Kovner, 1956).

3. *Evaporation and Transpiration as Related Processes*

The relative magnitudes of evaporation and transpiration depend largely upon the type of vegetation cover present but it has been suggested that the combined total water-loss of these two processes is not greatly influenced by the vegetation. Evaporation and transpiration are alike in needing a supply of energy to provide heat for the vaporization of water and in both cases incoming solar radiation is the main source of energy, although water vaporization may also be increased by drying winds. Once energy is used in either of the two processes it is no longer available, so that energy supply places an upper limit to the water loss from woodland ecosystems by evaporation and transpiration together. In Britain only about 40% of the incident radiation is used for the vaporization of water, energy being dissipated in various other ways, e.g. by reflection, re-radiation, heating the air, etc. Penman (1956) has stressed that the amount of energy available can be determined from meteorological data so that the potential loss of water from the ecosystem attributable to evaporation and transpiration can be calculated. He suggests, on theoretical grounds, that when there is adequate soil moisture and the vegetation forms a continuous cover, the total water loss due to evaporation and transpiration depends mainly on weather conditions and is not greatly influenced by differences in the type of vegetation cover. Considerable evidence has accumulated verifying the validity of Penman's conclusions (Zahner, 1955); at the same time it is recognized that Penman's qualifications (e.g. adequate water supply) are not always operative and that water loss is affected to a limited extent by differences in the vegetation cover because: (i) the amount of energy reflected depends on vegetation colour and the nature of its upper surface (Monteith, 1959); (ii) different plant covers vary in their resistance to wind; and (iii) the root systems of plants vary enormously, some taking water from a much bigger soil volume than others. Recently Rider (1957) has emphasized the need to apply cautiously the concept of equal potential transpiration irrespective of crop size.

4. *Run-off*

Water that is not evaporated or transpired may be retained within the ecosystem but the bulk passes from the ecosystem either as surface run-off over the mineral soil or as ground water through the soil. The relative proportions of the two types of water loss vary greatly, where surface run-off is excessive, flash floods and soil erosion occur commonly. Normally, more water leaves woodland ecosystems as subsurface than surface run-off. The organic matter overlying the mineral soil of woodlands tends to absorb water readily and Dunford (1954) found that when litter was removed from Ponderosa pine woodlands surface run-off increased almost seven-fold. In addition, water percolation into the mineral soil is facilitated by numerous old root channels and animal tunnels. The effect of the fauna on the water cycle is largely ignored but animal activity is extremely important in creating conditions favourable to rapid infiltration of water into the soil. The traditional role of forests in flood control by delaying and reducing the magnitude of peak run-off is generally recognized.

B. WATER BALANCE

Detailed studies of isolated factors of the hydrologic cycle are difficult to integrate into a composite picture. At present, the most complete data available have been obtained by lysimeter and water catchment investigations, but the interpretation and accuracy of results obtained by the two techniques have been subjects of controversy. Nevertheless, both types of installation have repeatedly given results which are broadly similar and consistent. In Table VIII, summarized water budgets obtained by lysimeter and water catchment studies are presented, these are based on the assumption that long-term changes in the water content of the ecosystem are insignificant. In all cases, comparative data for woodland and for neighbouring treeless areas, or for the same wooded area after tree felling are given.

The removal of water attributable to evaporation and transpiration in Table VIII varies from 42 to 72% of the precipitation in the wooded areas whilst in areas devoid of vegetation or with a predominantly grass cover the range is 23 to 69%. On the whole, in dense woodlands evapora-

Table VIII. *Annual Water Circulation*

(Weights — 10^4 kg per ha, Percentages — percentages of the precipitation)

LYSIMETER INSTALLATIONS

Location	Utah U.S.A.				Yorkshire England				Castricum Holland						San Dimas U.S.A.			
Vegetation type	Aspen		Bare of vegetation		Sitka spruce		Grassland		Black pine		Hardwoods		Bare sand		Pine		Bare soil	
	Weight	%	Weight	%	Weight	%	Weight	%	Weight	%	Weight	%	Weight	%	Weight	%	Weight	%
Input (precipitation)	1 340	100	1 340	100	984	100	1 136	100	892	100	892	100	892	100	554	100	554	100
Output (evaporation +transpiration)	568	42	361	27	711	72	717	63	621	70	456	51	209	23	379	68	204	37
Output (surface+ground run-off)	772	58	979	73	273	28	419	37	271	30	436	49	683	77	175	32	350	63
Reference	A		A		B		B		C		C		C		D		D	

CATCHMENT INSTALLATIONS

Location	Wagon Wheel U.S.A.				Coshocton U.S.A.				Coweeta U.S.A.				Sperbelgraben Switzerland		Rappengraben Switzerland	
Vegetation type	Mixed forest		Trees felled		Mixed forest		Grassland		Mixed hardwoods		Trees felled		Mixed forest		Grassland	
	Weight	%	Weight	%	Weight	%	Weight	%	Weight	%	Weight	%	Weight	%	Weight	%
Input (precipitation)	538	100	528	100	1 182	100	1 064	100	1 580	100	1 585	100	1 555	100	1 648	100
Output (evaporation +transpiration)	381	71	343	65	849	72	731	69	1 047	66	655	41	877	56	756	46
Output (surface+ground run-off)	157	29	185	35	333	28	333	31	533	34	930	59	678	44	892	54
Reference	E		E		F		F		G		G		H		H	

A. Croft and Monninger, 1953 B. Law, 1956 C. Wind, 1958 D. Zinke, 1959
E. Bates and Henry, 1928 F. Dreibelbis and Post, 1941 G. Hoover, 1944 H. Burger, 1954

tion slightly exceeds transpiration. At all eight locations, the removal of water by the combined effects of evaporation and transpiration is greatest and run-off is least where woodland conditions exist. That forested areas give lower yields of water for use by man, compared with other types of vegetation such as grassland, is now generally accepted. The greater water use of forests can be attributed to several factors, e.g. the large amount of interception and the extensive tree root system which in dry weather permits trees to draw upon water reserves deep in the soil. In countries such as Britain, where the provision of an adequate water supply is a serious problem, the policy of the afforestation of water catchments has been questioned in view of the possible reduction of water yield (Law, 1956) but it is important to take into account the effect of woodlands in regulating run-off and preventing erosion. The greatest recorded annual loss of water due to evaporation and transpiration occurs at Coweeta and would involve the expenditure of over 6×10^{12} cal per ha, over 150 times the energy released yearly by organic matter breakdown. At Coweeta and Castricum, the differences in run-off between tree-covered and treeless areas is such that the trees must be utilizing an additional 2×10^{12} cal per ha of radiation for evaporation and transpiration in an average year.

VI. Circulation of Chemical Elements

The interchange of chemicals between living organisms and between the physical and biological components of woodland ecosystems forms an extremely intricate system, essentially of a cyclic nature and broadly following the organic system. The magnitude and detailed pattern of circulation are also characteristic of the individual chemical element concerned, since elements vary in their availability to plants, and are absorbed selectively by woodland organisms and distributed unevenly through the bodies of plants and animals. The circulation does not form a closed system, the ecosystem capital changes as chemical elements are added to or removed from it in various natural and artificial ways, some of which are listed in Table IX.

The passage of chemical elements through woodland ecosystems is usually expressed on an annual basis, a legacy of the pioneer work done in temperate woodlands, where the regular autumn leaf fall dominates the system. In some ways this is unrealistic, since a particular molecule may take less or more than a year to pass through the circuit. For instance, it is possible that, nutrients present in the bud scales or the spring inflorescences are released by decomposition and re-absorbed by the trees during the summer months, to be shed in the same year as part of the autumn leaf fall. The rapidity with which elements can be moved in woodland ecosystems has been demonstrated by the use of

TABLE IX

Processes Changing the Capital of Chemical Elements in Woodland Ecosystems

Increase of capital	Decrease of capital
Planting of seedling trees	Harvesting of forest products
Movement of animals into ecosystem	Movement of animals out of ecosystem
Collection of plant matter, e.g. leaves blown by wind	Loss of plant matter, e.g. pollen, leaves blown by wind
Precipitation	Run-off
Trapping of dust, aerosols etc.	Erosion of soil
Application of fertilizers	Fire
Soil and rock weathering	Construction of drains to remove excess water
Fixation from atmosphere	
Enlarging effective soil volume, e.g. by draining and breaking up of compact soil horizons	

radioactive tracers. Radioactive phosphorus ^{32}P injected into the base of the trunk of a birch tree during sunny weather was well-distributed throughout the tree crown within 28 hours. Part of the ^{32}P could be washed out of the tree crowns or fall to the ground in litter and be taken up again by woodland plants within a period of days. Unpublished results of J. P. Witherspoon show that when 2 mc of caesium-134 tracer were introduced into the trunk of *Quercus alba*, after 165 days the ^{134}Cs had become widely redistributed in the ecosystem. About 47% was in the trunk, branches and large roots, 46% was in the tree leaves and the remaining 7% had been removed from the trees either by rain, animals or litter fall prior to the main leaf fall in autumn. Of the 7% of ^{134}Cs lost from the trees 5% was in the top 10 cm of soil, 0·8% was in the small rootlets, and the remaining 1·2% was about equally distributed between the understorey plants and litter with less than 0·1% in the animals. The preceding figures do not include losses to insects, but estimates based on leaf surface area consumed by insects suggest that several per cent may have passed through the animal food chain. Some of this would have been recovered in the litter, but a small fraction might have been dispersed by animal movements. Elements such as sodium and potassium are more mobile than caesium and probably short cycles of this type are very important in their circulation. In contrast, the rate of movement through woodland ecosystems of part of the chemical capital may be very slow, minerals incorporated in tree trunks not being released for over a century.

Our knowledge of the circulation of trace elements is very incomplete.

Lounamaa (1956) has reviewed the existing records for boron, chromium, manganese, cobalt, nickel, copper, zinc, gallium, yttrium, zirconium, molybdenum, silver, cadmium, tin and lead in various conifer and angiosperm trees as well as in the woodland understorey plants. He found little difference between evergreen softwoods and deciduous hardwoods with respect to their trace element content. In contrast, marked differences occur in the concentrations of the major elements and sufficient data are available to demonstrate the main features of the passage of sodium, potassium, calcium, magnesium, phosphorus and nitrogen through different types of woodland ecosystems.

A. WEIGHTS OF CHEMICAL ELEMENTS IN WOODLAND ECOSYSTEMS

In Tables X to XV the amounts and distribution of sodium, potassium, calcium, magnesium, phosphorus and nitrogen are given for some of the woodland ecosystems of Table I during the summer period, when the trees are in full leaf. At other times of the year the distribution would be very different. Many supplementary data are available showing the range of weights of various elements in individual components of woodland ecosystems, particularly of the litter layers (Lutz and Chandler, 1946) and of the ground flora (Scott, 1955). Few results have been published of the amounts of chemical elements contained in the trees and in the woodland fauna and microflora. Grimshaw *et al.* (1958) report the chemical content of the canopy invertebrates in pine woodlands to be, as g per ha, 2·2 for Na, 7·5 for K and 8·1 for P, whilst the birds of the three most common species contain 0·04 of Na, 0·06 of K, 0·32 of Ca, 0·01 of Mg and 0·2 of P g per ha. Consequently the omission of the fauna from Tables X to XV probably does not modify greatly the total figures. Part of the soil fauna and the soil microflora would be included in the litter layers.

The accumulation of mineral elements within the organic matter of the woodland ecosystems is greatest on the more fertile, nutrient rich soils. Nevertheless, woodland plants are able to regulate the concentration of elements within their bodies to some extent, so that differences in the nutrient contents of the plant organic mass of woodlands of the same tree species and of equal age but growing on very different soils are not so great as might be expected from differences in the availability of soil nutrients. In general, there is a fairly close relationship between the amount of organic matter present and the total amount of chemical elements it contains, although this relationship is affected by the nature of the organic matter; thus tree leaves, the ground flora and small twigs have much greater concentrations of chemicals than tree trunks. The tropical forest of Ghana with its huge weight of organic matter has the largest weights present with values for K, Ca, Mg, P and N of about 900, 2 700, 390, 150 and 2 050 kg per ha; in the temperate woodlands the maximum recorded values are Na 50, K 470, Ca 1 350, Mg 150, P 130 and N 1 150 per kg ha. The rate of build up of chemicals into the organic part of the ecosystem varies for different woodlands, the greatest mean annual accretion for the lifetime of the existing trees obtained from the tables being 1 kg per ha of Na, 23 for K, 53 for Ca, 8 for Mg, 6 for P, and 60 for N. Few data are available of the carbon contained within forest ecosystems but the organic mass of 350×10^3 kg per ha recorded for the 50-year-old woodlands gives a carbon content of about 175×10^3 kg per ha with an average annual accretion of 3 500 kg of C per ha.

The amounts of the various elements in the mineral soil given in the Tables are undoubtedly too small since it seems likely that the tree roots penetrate through a greater soil depth than has been sampled and it should be emphasized that except for N these are exchangeable, not total values. The amounts of exchangeable chemical elements present in the soil are not greatly in excess of the weights accumulated in the organic matter of woodland ecosystems.

B. REMOVAL OF CHEMICAL ELEMENTS BY HARVESTING OF TREE TRUNKS

Large volumes of timber have been extracted from some of the older woodlands in Tables X to XV. At each harvesting occasion considerable amounts of chemicals are removed from the ecosystem in the harvested tree trunks but the overall average annual loss is small. If the standing trees were felled and all boles harvested, the average annual loss, inclusive of past harvestings, would not exceed 0·2 kg per ha of Na, 7 of K, 15 of Ca, 2 of Mg, 14 of P and 14 of N. In fact, smaller amounts would normally be removed since the nutrient-rich upper parts of tree trunks and the bark are frequently left within the ecosystem and some tree boles are not harvested. The total weight of elements removed in the forest crop is relatively small compared with that of agriculture, Russell (1956) gives annual removal values for typical English crops of up to 99 kg per ha for Na, 279 for K, 72 for Ca, 29 for Mg, 26 for P and 167 for N. Similar figures for the annual nutrient uptake of field crops in Germany are given by Ehwald (1957), i.e. K 167, Ca 63, P 24 and N 164 kg per ha.

C. NUTRIENT UPTAKE, RETENTION AND RELEASE

The weight of chemical elements in the organic matter of woodland ecosystems does not represent the total uptake by the plants since, apart from losses due to harvesting, chemical elements are continually

Table X

Sodium (kg per ha) *in Plant and Soil Material in Ecosystems*

Trees	*Pinus sylvestris*	*Pinus sylvestris*	*Pinus sylvestris*	*Pinus sylvestris*	*Pinus sylvestris*	*Pinus nigra*	*Pinus nigra*	*Picea abies*	*Picea abies*	*Picea abies*	*Picea abies*	*Picea abies*
Location	England	Scotland	England	England	Scotland	England	Scotland	England	England	England	Sweden	Sweden
Stand Number	1	4	5	6	7	10	11	12	18	19	20	21
Age of trees in years	23	33	47	55	64	46	48	20	47	47	52	58
Living plants												
Tree layer												
Fruit	<1	—	}	<1	—	}	—	}	}	}	—	—
Leaves	1	3	} 4	1	—	} 2	—	} 12	} 3	} 9	—	—
Branches	1	2	}	1	—	}	—	}	}	}	—	—
Trunks	1	<1	2	5	—	3	—	2	2	3	—	—
Understorey plants	<1	<1	2	4	—	3	—	0	0	0	—	—
Roots	6	—	—	30	—	—	—	—	—	—	—	—
Dead plant material												
On trees (branches)	<1	1	—	1	—	—	—	—	—	—	—	—
Litter on ground (L, F and H)	3	34	2	6	—	3	—	3	4	5	—	—
Mineral soil (exchangeable)	—	—	57	—	—	79	—	68	85	72	—	—
Removed in harvested trunks	<1	0	3	5	—	2	—	<1	3	3	—	—
Total removed in tree trunks if clear felled	1	<1	5	10	—	5	—	2	5	6	—	—
Depth of sampling in mineral soil cm	—	—	70	—	—	70	—	60	70	70	—	—
Reference	A	B	C	A	D	C	D	C	C	C	E	E

For key to references see foot of Table XV.

TABLE X. — *continued*

Trees	*Pseudotsuga taxifolia*	*Pseudotsuga taxifolia*	*Pseudotsuga taxifolia*	*Pseudotsuga taxifolia*	*Pseudotsuga taxifolia*	*Pseudotsuga taxifolia*	*Pseudotsuga taxifolia*	*Pseudotsuga taxifolia*	*Larix decidua*
Location	England	England	U.S.A.	U.S.A.	U.S.A.	U.S.A.	England	U.S.A.	England
Stand Number	24	25	26	27	28	29	30	31	32
Age of trees in years	21	22	29	32	38	38	47	52	46
Living plants									
Tree layer									
Fruit			—	—	—	—		—	
Leaves	} 8	} 3	—	—	—	—	} 6	—	} 4
Branches			—	—	—	—		—	
Trunks	2	2	—	—	—	—	4	—	3
Understorey plants	<1	0	—	—	—	—	2	—	5
Roots	—	—	—	—	—	—	—	—	—
Dead plant material									
On trees (branches)	—	—	—	—	—	—	—	—	—
Litter on ground (L, F and H)	3	6	—	—	—	—	1	—	4
Mineral soil (exchangeable)	101	105	—	—	—	—	107	—	64
Removed in harvested trunks	1	<1	—	—	—	—	4	—	2
Total removed in tree trunks if clear felled	3	2	—	—	—	—	8	—	5
Depth of sampling in mineral soil cm	60	70	—	—	—	—	70	—	70
Reference	C	C	F	F	F	F	C	F	C

TABLE X. — *continued*

Trees	*Betula verrucosa*	*Betula verrucosa*	*Betula verrucosa*	*Alnus incana*	*Quercus petraea*	*Quercus robur*	*Fagus sylvatica*	*Nothofagus obliqua*	*Nothofagus truncata*	*Castanea sativa*	Mixed tropical
Location	England	England	England	England	England	England	England	England	New Zealand	England	Ghana
Stand Number	34	35	38	43	44	47	51	54	55	56	60
Age of trees in years	22	24	55	22	21	47	39	22	110	47	50
Living plants											
Tree layer											
Fruit	}	—	—	}	}	}	}	}	—	}	—
Leaves	} 1	1	1	} 4	} 3	} 1	} 1	} 16	4	} 1	—
Branches	}	1	3	}	}	}	}	}	7	}	—
Trunks	1	1	2	3	1	3	2	2	9	2	—
Understorey plants	1	—	—	1	1	1	1	2	—	1	—
Roots	—	1	4	—	—	—	—	—	3	—	—
Dead plant material											
On trees (branches)	—	1	1	—	—	—	—	—	1	—	—
Litter on ground (L, F and H)	1	—	—	1	1	1	1	1	8	1	—
Mineral soil (exchangeable)	88	—	—	87	54	70	47	106	—	59	—
Removed in harvested trunks	0	0	0	1	0	1	1	1	—	1	—
Total removed in tree trunks if clear felled	1	1	2	4	1	4	3	3	—	3	—
Depth of sampling in mineral soil cm	70	—	—	70	60	70	70	60	—	70	30
Reference	C	G	G	C	C	C	C	C	H	C	I

Table XI

Potassium (kg per ha) *in Plant and Soil Material in Ecosystems*

Trees	*Pinus sylvestris*	*Pinus sylvestris*	*Pinus sylvestris*	*Pinus sylvestris*	*Pinus sylvestris*	*Pinus nigra*	*Pinus nigra*	*Picea abies*	*Picea abies*	*Picea abies*	*Picea abies*	*Picea abies*
Location	England	Scotland	England	England	Scotland	England	Scotland	England	England	England	Sweden	Sweden
Stand Number	3	4	5	6	7	10	11	12	18	19	20	21
Age of trees in years	23	33	47	55	64	46	48	20	47	47	52	58
Living plants												
Tree layer												
Fruit	1	—		2	—		—				—	—
Leaves	52	43	}95	40	19	}93	36	}301	}117	}166	49	56
Branches	28	39		19	29		29				27	29
Trunks	34	84	54	46	51	102	68	136	44	60	65	52
Understorey plants	2	<1	165	22	—	156	—	0	0	0	—	—
Roots	35	54	—	43	—	—	—	—	—	—	—	—
Dead plant material												
On trees (branches)	4	4	—	2	—	—	—	—	—	—	1	2
Litter on ground (L, F and and H)	24	165	11	34	—	24	—	31	30	21	—	273†
Mineral soil (exchangeable)	—	—	249	—	—	359	—	147	238	233	—	—
Removed in harvested trunks	4	0	88	98	—	89	—	6	85	59	—	—
Total removed in tree trunks if clear felled	38	84	142	144	—	191	68	142	129	119	65	52
Depth of sampling in mineral soil cm	—	—	70	—	—	70	—	60	70	70	—	—
Reference	A	B	C	A	D	C	D	C	C	C	E	E

† Includes Ao + A1 soil horizon.
For key to references see foot of Table XV.

TABLE XI.—*continued*

Trees	*Pseudotsuga taxifolia*	*Pseudotsuga taxifolia*	*Pseudotsuga taxifolia*	*Pseudotsuga taxifolia*	*Pseudotsuga taxifolia*	*Pseudotsuga taxifolia*	*Pseudotsuga taxifolia*	*Pseudotsuga taxifolia*	*Larix decidua*
Location	England	England	U.S.A.	U.S.A.	U.S.A.	U.S.A.	England	U.S.A.	England
Stand Number	24	25	26	27	28	29	30	31	32
Age of trees in years	21	22	29	32	38	38	47	52	46
Living plants									
Tree layer									
Fruit, Leaves, Branches	68	241	94	54	131	138	136	172	110
Trunks	63	45	—	—	—	—	73	—	32
Understorey plants	5	0	66	22	10	8	39	—	74
Roots	—	—	86	55	28	43	—	21	—
Dead plant material									
On trees (branches)	—	—	—	—	—	—	—	—	—
Litter on ground (L, F and H)	25	18	—	—	—	—	9	—	23
Mineral soil (exchangeable)	199	87	—	—	—	—	253	—	223
Removed in harvested trunks	25	11	—	—	—	—	75	—	19
Total removed in tree trunks if clear felled	88	56	—	—	—	—	148	—	51
Depth of sampling in mineral soil cm	60	70	—	—	—	—	70	—	70
Reference	C	C	F	F	F	F	C	F	C

TABLE XI.—*continued*

Trees	*Betula verrucosa*	*Betula verrucosa*	*Betula verrucosa*	*Alnus incana*	*Quercus petraea*	*Quercus robur*	*Fagus sylvatica*	*Nothofagus obliqua*	*Nothofagus truncata*	*Castanea sativa*	Mixed tropical
Location	England	England	England	England	England	England	England	England	New Zealand	England	Ghana
Stand Number	34	35	38	43	44	47	51	54	55	56	60
Age of trees in years	22	24	55	22	21	47	39	22	110	47	50
Living plants											
Tree layer											
Fruit		—	—						—		
Leaves	45	28	43	78	61	105	91	141	13	37	
Branches		21	46						119		808
Trunks	17	29	65	50	41	118	94	76	250	31	
Understorey plants	14	—	—	22	14	23	2	13	—	9	
Roots	—	15	46	—	—	—	—	—	67	—	87
Dead plant material											
On trees (branches)	—	<1	<1	—	—	—	—	—	<1	—	—
Litter on ground (L, F and H	8	—	—	5	11	8	20	7	15	10	10
Mineral soil (exchangeable)	125	—	—	119	141	329	254	156	—	239	649
Removed in harvested trunks	0	0	0	8	0	63	61	22	—	17	—
Total removed in tree trunks if clear felled	17	29	65	58	41	181	155	98	—	48	—
Depth of sampling in mineral soil cm	70	—	—	70	60	70	70	60	—	70	30
Reference	C	G	G	C	C	C	C	C	H	C	I

Table XII

Calcium (kg per ha) *in Plant and Soil Material in Ecosystems*

Trees	*Pinus sylvestris*	*Pinus sylvestris*	*Pinus sylvestris*	*Pinus sylvestris*	*Pinus sylvestris*	*Pinus nigra*	*Pinus nigra*	*Picea abies*	*Picea abies*	*Picea abies*	*Picea abies*	*Picea abies*
Location	England	Scotland	England	England	Scotland	England	Scotland	England	England	England	Sweden	Sweden
Stand Number	3	4	5	6	7	10	11	12	18	19	20	21
Age of trees in years	23	33	47	55	64	46	48	20	47	47	52	58
Living plants												
Tree layer												
Fruit	<1	—		<1	—		—				—	—
Leaves	36	36	44	34	22	75	27	167	105	309	50	76
Branches	34	30		23	35		50				44	63
Trunks	56	115	119	122	143	115	79	190	107	198	108	120
Understorey plants	1	<1	41	19	—	55	—	0	0	0	—	—
Roots	48	33	—	93	—	—	—	—	—	—	—	—
Dead plant material												
On trees (branches)	15	13	—	9	—	—	—	—	—	—	5	11
Litter on ground (L, F and H)	139	322	25	141	—	67	—	73	65	117	—	265†
Mineral soil (exchangeable)	—	—	437	—	—	473	—	293	458	475	—	—
Removed in harvested trunks	6	0	196	210	—	100	—	8	209	194	—	—
Total removed in tree trunks if clear felled	62	115	315	332	—	215	79	198	316	392	108	120
Depth of sampling in mineral soil cm	—	—	70	—	—	70	—	60	70	70	—	—
Reference	A	B	C	A	D	C	D	C	C	C	E	E

† Includes Ao + A1 soil horizon.
For key to references see foot of Table XV.

Table XII. — *continued*

Trees	*Pseudotsuga taxifolia*	*Pseudotsuga taxifolia*	*Pseudotsuga taxifolia*	*Pseudotsuga taxifolia*	*Pseudotsuga taxifolia*	*Pseudotsuga taxifolia*	*Pseudotsuga taxifolia*	*Pseudotsuga taxifolia*	*Larix decidua*
Location	England	England	U.S.A.	U.S.A.	U.S.A.	U.S.A.	England	U.S.A.	England
Stand Number	24	25	26	27	28	29	30	31	32
Age of trees in years	21	22	29	32	38	38	47	52	46
Living plants									
Tree layer									
Fruit			—	—	—	—		—	
Leaves	}98	}557	—	—	—	—	}246	—	}100
Branches			—	—	—	—		—	
Trunks	54	117	—	—	—	—	92	—	72
Understorey plants	1	0	—	—	—	—	23	—	26
Roots	—	—	—	—	—	—	—	—	—
Dead plant material									
On trees (branches)	—	—	—	—	—	—	—	—	—
Litter on ground (L, F and H)	68	230	—	—	—	—	27	—	104
Mineral soil (exchangeable)	475	6 240	—	—	—	—	534	—	404
Removed in harvested trunks	29	28	—	—	—	—	94	—	42
Total removed in tree trunks if clear felled	83	145	—	—	—	—	186	—	114
Depth of sampling in mineral soil cm	60	70	—	—	—	—	70	—	70
Reference	C	C	F	F	F	F	C	F	C

Table XII. — *continued*

Trees	*Betula verrucosa*	*Betula verrucosa*	*Betula verrucosa*	*Alnus incana*	*Quercus petraea*	*Quercus robur*	*Fagus sylvatica*	*Nothofagus obliqua*	*Nothofagus truncata*	*Castanea sativa*	Mixed tropical
Location	England	England	England	England	England	England	England	England	New Zealand	England	Ghana
Stand Number	34	35	38	43	44	47	51	54	55	56	60
Age of trees in years	22	24	55	22	21	47	39	22	110	47	50
Living plants											
Tree layer											
Fruit		—	—						—		
Leaves	}192	32	42	}304	}61	}73	}72	}167	20	}41	
Branches		69	180						281		}2 477
Trunks	120	132	275	280	58	173	79	105	647	106	
Understorey plants	15	—	—	46	4	11	<1	5	—	7	
Roots	—	79	154	—	—	—	—	—	172	—	145
Dead plant material											
On trees (branches)	—	9	10	—	—	—	—	—	9	—	—
Litter on ground (L, F and H)	72	—	—	72	36	35	51	39	215	32	45
Mineral soil (exchangeable)	7 105	—	—	8 260	269	424	365	344	—	424	2 573
Removed in harvested trunks	0	0	0	47	0	92	51	30	—	58	—
Total removed in tree trunks if clear felled	120	132	275	327	58	265	130	135	—	164	—
Depth of sampling in mineral soil cm	70	—	—	70	60	70	70	60	—	70	30
Reference	C	G	G	C	C	C	C	C	H	C	I

Table XIII

Magnesium (kg per ha) *in Plant and Soil Material in Ecosystems*

Trees	*Pinus sylvestris*	*Pinus sylvestris*	*Pinus sylvestris*	*Pinus sylvestris*	*Pinus sylvestris*	*Pinus nigra*	*Pinus nigra*	*Picea abies*	*Picea abies*	*Picea abies*	*Picea abies*	*Picea abies*
Location	England	Scotland	England	England	Scotland	England	Scotland	England	England	England	Sweden	Sweden
Stand Number	3	4	5	6	7	10	11	12	18	19	20	21
Age of trees in years	23	33	47	55	64	46	48	20	47	47	52	58
Living plants												
Tree layer												
Fruit	<1	—		<1	—		—				—	—
Leaves	8	6	6	8	4	14	7	48	20	50	9	—
Branches	10	7		8	8		10				8	—
Trunks	8	24	25	28	23	39	23	34	19	35	18	—
Understorey plants	<1	<1	12	5	—	10	—	0	0	0	—	—
Roots	16	13	—	20	—	—	—	—	—	—	—	—
Dead plant material												
On trees (branches)	2	2	—	4	—	—	—	—	—	—	1	—
Litter on ground (L, F and H)	17	31	9	31	—	15	—	23	27	19	—	—
Mineral soil (exchangeable)	—	—	299	—	—	269	—	227	300	328	—	—
Removed in harvested trunks	3	0	41	42	—	34	—	1	37	34	—	—
Total removed in tree trunks if clear felled	11	24	66	70	—	73	—	35	56	69	18	—
Depth of sampling in mineral soil cm	—	—	70	—	—	70	—	60	70	70	—	—
Reference	A	B	C	A	D	C	D	C	C	C	E	E

For key to references see foot of Table XV.

TABLE XIII. — *continued*

Trees	*Pseudotsuga taxifolia*	*Pseudotsuga taxifolia*	*Pseudotsuga taxifolia*	*Pseudotsuga taxifolia*	*Pseudotsuga taxifolia*	*Pseudotsuga taxifolia*	*Pseudotsuga taxifolia*	*Pseudotsuga taxifolia*	*Larix decidua*
Location	England	England	U.S.A.	U.S.A.	U.S.A.	U.S.A.	England	U.S.A.	England
Stand Number	24	25	26	27	28	29	30	31	32
Age of trees in years	21	22	29	32	38	38	47	52	46
Living plants									
Tree layer									
Fruit			—	—	—	—		—	
Leaves	18 (fruit, leaves and branches)	39 (fruit, leaves and branches)	—	—	—	—	38 (fruit, leaves and branches)	—	14 (fruit, leaves and branches)
Branches			—	—	—	—		—	
Trunks	11	10	—	—	—	—	24	—	21
Understorey plants	1	0	—	—	—	—	5	—	8
Roots	—	—	—	—	—	—	—	—	—
Dead plant material									
On trees (branches)	—	—	—	—	—	—	—	—	—
Litter on ground (L, F and H)	9	9	—	—	—	—	7	—	27
Mineral soil (exchangeable)	270	250	—	—	—	—	285	—	254
Removed in harvested trunks	4	2	—	—	—	—	24	—	12
Total removed in tree trunks if clear felled	15	12	—	—	—	—	48	—	33
Depth of sampling in mineral soil cm	60	70	—	—	—	—	70	—	70
Reference	C	C	F	F	F	F	C	F	C

Table XIII. — *continued*

Trees	*Betula verrucosa*	*Betula verrucosa*	*Betula verrucosa*	*Alnus incana*	*Quercus petraea*	*Quercus robur*	*Fagus sylvatica*	*Nothofagus obliqua*	*Nothofagus truncata*	*Castanea sativa*	Mixed tropical
Location	England	England	England	England	England	England	England	England	N϶w Zealand	England	Ghana
Stand Number	34	35	38	43	44	47	51	54	55	56	60
Age of trees in years	22	24	55	22	21	47	39	22	110	47	50
Living plants											
Tree layer											
Fruit		—	—						—		
Leaves	}17	5	7	}23	}17	}18	}14	}25	4	}7	
Branches		6	13						40		}340
Trunks	7	13	26	16	11	23	28	12	58	20	
Understorey plants	2	—	—	5	2	4	<1	2	—	3	
Roots	—	4	14	—	—	—	—	—	21	—	44
Dead plant material											
On trees (branches)	—	<1	<1	—	—	—	—	—	1	—	—
Litter on ground (L, F and H)	4	—	—	2	7	5	14	5	23	7	6
Mineral soil (exchangeable)	156	—	—	213	192	301	249	278	—	266	369
Removed in harvested trunks	0	0	0	3	0	12	18	3	—	13	—
Total removed in tree trunks if clear felled	7	13	26	19	11	35	46	15	—	33	—
Depth of sampling in mineral soil cm	70	—	—	70	60	70	70	60	—	70	30
Reference	C	G	G	C	C	C	C	C	H	C	I

Table XIV

Phosphorus (kg per ha) *in Plant and Soil Material in Ecosystems*

Trees	*Pinus sylvestris*	*Pinus sylvestris*	*Pinus sylvestris*	*Pinus sylvestris*	*Pinus sylvestris*	*Pinus nigra*	*Pinus nigra*	*Picea abies*	*Picea abies*	*Picea abies*	*Picea abies*	*Picea abies*
Location	England	Scotland	England	England	Scotland	England	Scotland	England	England	England	Sweden	Sweden
Stand Number	3	4	5	6	7	10	11	12	18	19	20	21
Age of trees in years	23	33	47	55	64	46	48	20	47	47	52	58
Living plants												
Tree layer												
Fruit	<1	—		<1	—		—				—	—
Leaves	10	9	17	11	5	17	7	65	26	64	20	15
Branches	5	7		6	7		5				9	8
Trunks	5	12	8	7	11	14	11	34	11	18	12	8
Understorey plants	<1	<1	9	3	—	7	—	0	0	0	—	—
Roots	11	11	—	17	—	—	—	—	—	—	—	—
Dead plant material												
On trees (branches)	3	2	—	1	—	—	—	—	—	—	1	1
Litter on ground (L, F and H)	22	78	9	28	—	15	—	20	19	24	—	155†
Mineral soil (exchangeable)	—	—	107	—	—	59	—	5	102	319	—	—
Removed in harvested trunks	2	0	13	14	—	12	—	1	22	18	—	—
Total removed in tree trunks if clear felled	7	12	21	21	—	26	11	35	33	39	12	8
Depth of sampling in mineral soil cm	—	—	70	—	—	70	—	60	70	70	—	—
Reference	A	B	C	A	D	C	D	C	C	C	E	E

† Includes Ao + A1 soil horizon.
For key to references see foot of Table XV.

Table XIV. — *continued*

Trees	*Pseudotsuga taxifolia*	*Pseudotsuga taxifolia*	*Pseudotsuga taxifolia*	*Pseudotsuga taxifolia*	*Pseudotsuga taxifolia*	*Pseudotsuga taxifolia*	*Pseudotsuga taxifolia*	*Pseudotsuga taxifolia*	*Larix decidua*
Location	England	England	U.S.A.	U.S.A.	U.S.A.	U.S.A.	England	U.S.A.	England
Stand Number	24	25	26	27	28	29	30	31	32
Age of trees in years	21	22	29	32	38	38	47	52	46
Living plants									
Tree layer									
Fruit									
Leaves	}13	}94					}34		}31
Branches			}30	}26	}45	}50		}72	
Trunks	8	14					15		13
Understorey plants	<1	0	8	3	1	1	6	—	10
Roots	—	—	18	13	7	12	—	11	—
Dead plant material									
On trees (branches)	—	—	—	—	—	—	—	—	—
Litter on ground (L, F and H)	11	22	—	—	—	—	7	—	25
Mineral soil (exchangeable)	4	204	—	—	—	—	50	—	463
Removed in harvested trunks	3	3	—	—	—	—	16	—	8
Total removed in tree trunks if clear felled	11	17	—	—	—	—	31	—	21
Depth of sampling in mineral soil cm	60	70	—	—	—	—	70	—	70
Reference	C	C	F	F	F	F	C	F	C

TABLE XIV. — *continued*

Trees	*Betula verrucosa*	*Betula verrucosa*	*Betula verrucosa*	*Alnus incana*	*Quercus petraea*	*Quercus robur*	*Fagus sylvatica*	*Nothofagus obliqua*	*Nothofagus truncata*	*Castanea sativa*	Mixed tropical
Location	England	England	England	England	England	England	England	England	New Zealand	England	Ghana
Stand Number	34	35	38	43	44	47	51	54	55	56	60
Age of trees in years	22	24	55	22	21	47	39	22	110	47	50
Living plants											
Tree layer											
Fruit		—	—						—		
Leaves	}22	3	4	}30	}13	}21	}22	}23	2	}6	
Branches		5	12						19		}124
Trunks	8	6	11	26	6	11	16	10	32	6	
Understorey plants	3	—	—	4	1	3	<1	1	—	2	
Roots	—	5	7	—	—	—	—	—	21	—	11
Dead plant material											
On trees (branches)	—	<1	<1	—	—	—	—	—	<1	—	—
Litter on ground (L, F and H)	6	—	—	5	6	5	11	5	7	6	1
Mineral soil (exchangeable)	230	—	—	194	4	36	42	3	—	39	13
Removed in harvested trunks	0	0	0	4	0	6	11	3	—	3	—
Total removed in tree trunks if clear felled	8	6	11	30	6	17	27	13	—	9	—
Depth of sampling in mineral soil cm	70	—	—	70	60	70	70	60	—	70	30
Reference	C	G	G	C	C	C	C	C	H	C	I

TABLE XV

Nitrogen (kg per ha) *in Plant and Soil Material in Ecosystems*

Trees	*Pinus sylvestris*	*Pinus sylvestris*	*Pinus sylvestris*	*Pinus sylvestris*	*Pinus sylvestris*	*Pinus nigra*	*Pinus nigra*	*Picea abies*	*Picea abies*	*Picea abies*	*Picea abies*	*Picea abies*
Location	England	Scotland	England	England	Scotland	England	Scotland	England	England	England	Sweden	Sweden
Stand Number	3	4	5	6	7	10	11	12	18	19	20	21
Age of trees in years	23	33	47	55	64	46	48	20	47	47	52	58
Living plants												
Tree layer												
Fruit	1	—	}	<1	—	}	—	}	}	}	—	—
Leaves	115	89	}179	124	51	}207	52	}599	}226	}573	128	91
Branches	60	52	}	57	53	}	35	}	}	}	74	60
Trunks	59	97	90	88	101	229	98	260	105	132	109	63
Understorey plants	2	<1	95	40	—	87	—	0	0	0	—	—
Roots	189	81	—	184	—	—	—	—	—	—	—	—
Dead plant material												
On trees (branches)	44	27	—	—	—	—	—	—	—	—	8	9
Litter on ground (L, F and H)	163	1 622	192	409	—	274	—	294	330	436	—	1 570†
Mineral soil	—	—	7 308	—	—	6 619	—	4 740	6 596	7 781	—	—
Removed in harvested trunks	7	0	148	161	—	199	—	11	205	129	—	—
Total removed in tree trunks if clear felled	66	97	238	249	—	428	—	271	310	261	109	63
Depth of sampling in mineral soil cm	—	—	70	—	—	70	—	60	70	70	—	—
Reference	A	B	C	A	D	C	D	C	C	C	E	E

† Includes Ao + Al soil horizon.

A. Ovington, 1959a, b
B. Ovington and Madgwick, 1959a
C. Ovington, 1954a; 1955; 1956a, b, c; 1957a, b; 1958a, b, c; 1959c
D. Wright and Will, 1958
E. Tamm and Carbonnier, 1961
F. Heilman, 1961
G. Ovington and Madgwick, 1959c
H. Miller, in manuscript
I. Greenland and Kowal, 1960
J. Ogawa *et al.*, 1961.

TABLE XV. — *continued*

Trees	*Pseudotsuga taxifolia*	*Pseudotsuga taxifolia*	*Pseudotsuga taxifolia*	*Pseudotsuga taxifolia*	*Pseudotsuga taxifolia*	*Pseudotsuga taxifolia*	*Pseudotsuga taxifolia*	*Pseudotsuga taxifolia*	*Larix decidua*
Location	England	England	U.S.A.	U.S.A.	U.S.A.	U.S.A.	England	U.S.A.	England
Stand Number	24	25	26	27	28	29	30	31	32
Age of trees in years	21	22	29	32	38	38	47	52	46
Living plants									
Tree layer									
Fruit									
Leaves	}213	}704					}311		}357
Branches			}112	}84	}196	}260		}360	
Trunks	99	109					157		108
Understorey plants	4	0	59	21	10	10	49	—	60
Roots	—	—	90	68	35	67	—	49	—
Dead plant material									
On trees (branches)	—	—	—	—	—	—	—	—	—
Litter on ground (L, F and H)	156	188	—	—	—	—	119	—	594
Mineral soil	5 805	1 639	—	—	—	—	7 701	—	6 125
Removed in harvested trunks	40	26	—	—	—	—	160	—	63
Total removed in tree trunks if clear felled	139	135	—	—	—	—	317	—	171
Depth of sampling in mineral soil cm	60	70	—	—	—	—	70	—	70
Reference	C	C	F	F	F	F	C	F	C

Table XV. — *continued*

Trees	*Betula verrucosa*	*Betula verrucosa*	*Betula verrucosa*	*Alnus incana*	*Quercus petraea*	*Quercus robur*	*Fagus sylvatica*	*Nothofagus obliqua*	*Nothofagus truncata*	*Castanea sativa*	Mixed tropical
Location	England	England	England	England	England	England	England	England	New Zealand	England	Ghana
Stand Number	34	35	38	43	44	47	51	54	55	56	60
Age of trees in years	22	24	55	22	21	47	39	22	110	47	50
Living plants											
Tree layer											
Fruit		—	—						—		
Leaves	178	47	78	420	214	218	155	404	32	65	
Branches		68	168						137		1 794
Trunks	62	73	145	220	70	151	128	105	172	116	
Understorey plants	24	—	—	50	11	24	2	11	—	15	
Roots	—	89	152	—	—	—	—	—	65	—	214
Dead plant material											
On trees (branches)	—	7	3	—	—	—	—	—	2	—	—
Litter on ground (L, F and H)	47	—	—	65	121	71	180	71	125	80	35
Mineral soil	1 300	—	—	2 396	5 103	7 476	6 640	5 780	—	6 863	4 587
Removed in harvested trunks	0	0	0	37	0	80	83	30	—	64	—
Total removed in tree trunks if clear felled	62	73	145	257	70	231	211	135	—	180	—
Depth of sampling in mineral soil cm	70	—	—	70	60	70	70	60	—	70	30
Reference	C	G	G	C	C	C	C	C	H	C	I

TABLE XV. — *continued*

Trees	Dipterocarps	Mixed savanna	Temperate savanna	Evergreen gallery
Location	Thailand	Thailand	Thailand	Thailand
Stand Number	61	62	63	64
Age of trees in years	—	—	—	—
Living plants				
Tree layer				
Fruit	—	—	—	—
Leaves	136	133	400	526
Branches, Trunks	81	97	282	412
Understorey plants	18	27	3	2
Roots	83	92	205	282
Dead plant material				
On trees (branches)	—	—	—	—
Litter on ground (L, F and H)	46	38	711	39
Mineral soil (exchangeable)	538	538	5 324	853
Depth of sampling in mineral soil cm	20	20	20	20
Reference	J	J	J	J

being released by the decomposition of organic matter. Russian scientists have been particularly interested in the determination of nutrient uptake, retention and release by trees and some typical results are given in Table XVI. The large differences between nutrient uptake and accretion are probably underestimates since in measuring uptake, usually no account is taken of nutrients leached out of leaves, or root exudates (Vinokurov and Tyurmenko, 1960), and of the annual mortality of fine roots. As a woodland matures, the magnitude of nutrient exchange between the trees and soil changes and is greatest about the time when organic matter production is at a maximum. Smirnova and Gorodentseva (1958) found the annual uptake, retention and release of nitrogen by the trees in a 30-year-old birch wood to be 111, 45 and 66 kg per ha respectively, whilst the corresponding values in a 67-year-old birch wood are 51, 3 and 48. In young birchwoods from 30 to 50% of the nutrient uptake is retained in the tree biomass whilst in old birchwoods only 10 to 20% is so immobilized. Mina (1955) reports that in mixed oakwoods of the Russian forest steppe region, uptake and retention of nutrients reach a maximum when the trees are about 50 years of age. Vinokurov and Tyurmenko (1958) have analysed forest soils throughout the year and suggest that nutrient uptake varies seasonally, reaching a peak at the end of June or beginning of July.

The part played by understorey vegetation in nutrient circulation must not be overlooked since, at certain stages in the life history of a woodland, a greater weight of chemical elements may be circulating through the understorey than through the trees. P'Yavchenko (1960) found this to be the case in old spruce and pine forests in the Vologod region of U.S.S.R. In an 120-year-old pine forest the annual production of dry matter by the understorey vegetation was sixteen times greater than that by the trees and the relative amounts of nutrients absorbed were even greater. A comparable situation occurs in young and old forest plantations in Britain where the ground flora is very luxuriant.

Nutrient uptake by woodland plants and nutrient release by decomposition are rarely in balance so that changes in the nutrient capital of the organic matter of the ecosystem must be taken into account in preparing balance tables (Table XVII). Trees, established in formerly treeless areas, represent a positive gain but long-term changes in the nutrient content of the understorey vegetation and litter layers may be either positive or negative. Finally, changes in the exchangeable and total amounts of nutrients in the mineral soil resulting from woodland growth, may be very important.

TABLE XVI

Annual Uptake, Retention and Release of Nutrients by the Trees of Woodland Ecosystems

kg per ha

		Chemical Elements (kg per ha)					Reference
		K	Ca	Mg	P	N	
Spruce on podsol soil U.S.S.R.	Uptake	18	56	9	3·2	61	Sonn, 1960
	Retained	6	8	2	0·6	9	
	Released by litter fall	12	48	7	2·6	52	
Spruce on rich soil U.S.S.R.	Uptake	20	52	7	2·7	62	Sonn, 1960
	Retained	3	6	1	0·3	8	
	Released by litter fall	17	46	6	2·4	54	
Birch U.S.S.R.	Uptake	30	107	29	11	111	Smirnova and Gorodentseva, 1958
	Retained	17	53	10	6	45	
	Released by litter fall	13	54	19	5	66	
Mixed oakwood U.S.S.R.	Uptake	85	102	16	7	92	Mina, 1955
	Retained	23	16	3	4	33	
	Released by litter fall	62	86	13	3	59	
Nothofagus truncata New Zealand	Uptake	34	84	12	3·3	40	Miller, in manuscript
	Retained	4	10	1	0·7	3	
	Released by litter fall	30	74	11	2·6	37	

TABLE XVII

Balance Sheet for Plantations of Pinus sylvestris *for the 55 years from the time of planting*

	Na	K	Ca	Mg	P	N
			(kg per ha)			
Trees						
Uptake by trees	132	1 933	2 272	431	413	4 817
Present in living trees	36	150	272	64	41	453
Removed in harvested tree trunks	5	98	210	42	14	161
Left in crowns and roots of harvested trees	48	279	386	80	75	704
Litter fall from trees (leaves, branches, cones, etc.)	43	1 406	1 404	245	283	3 499
Transfer from trees to soil	91	1 685	1 790	325	358	4 203
Ground flora						
Uptake by ground flora	72	876	771	169	182	2 058
Present in ground flora	4	22	19	5	3	40
Change in ground flora	+2	−1	−4	−1	−5	−73
Litter fall from ground flora	70	877	775	170	187	2 131
Litter						
Total return from trees and ground flora	161	2 562	2 565	495	545	6 334
Present in litter layers (L, F and H)	6	34	141	30	28	409
Change in litter layers (L, F and H)	+6	+24	+115	+22	+19	+250
Released by decomposition	161	2 538	2 450	473	526	6 084
Average annual uptake by trees and ground flora	3·7	51	55	11	11	125
Average annual change in ecosystem	+0·8	+3	+7	+2	+1	+11
Average annual removal in tree trunks	<0·1	2	4	<1	0·3	3
Average annual release by decomposition	2·9	46	44	9	10	111

D. SOIL CHANGES

Whilst the average annual removal of nutrients in the forest crop is small compared with that under most agricultural systems, the long-term loss from woodlands by harvesting is considerable. Furthermore, the organic matter accumulated in woodlands contains large amounts of nutrients which are transferred from the soil and immobilized in the organic material. Clearly, if the soil-nutrient reserve is not supplemented in some way, a gradual depletion of the soil capital would occur as a woodland matures and is harvested. Since woodlands are frequently restricted to areas having soils of low fertility, serious soil degradation could result within a few generations of trees and reduce productivity (Rennie, 1955).

Numerous instances of change in the chemical and physical properties of woodland soils have been recorded (Day, 1940; Zinecker, 1950; Dimbleby, 1952; Giulimondi *et al.*, 1956; Duchaufour and Guinaudeau, 1957; Wittich, 1961). Generally, comparisons are made of the soils of woodlands of different tree species and of wooded and treeless areas established in close proximity on originally similar soils. The published results show a very wide range of rates of change and in some cases it appears that soil changes attributed to the growth of trees may in reality be due to initial differences in the sites being compared. The most obvious and well-documented pedological differences recorded are in the type, weight, percentage and total chemical compositions of the litter layers formed over the surface of the mineral soil. Changes in the underlying mineral soil resulting from woodland growth are more difficult to determine precisely and vary according to the inherent soil properties. For example, in nutrient-deficient acid soils or base-rich alkaline soils the soil pH is not likely to be changed greatly by a single generation of trees. Whilst there is evidence that woodlands may cause a redistribution of chemical elements within the soil profile, few data are available of changes in the weights of nutrients in the whole soil mass. An increase in the amount of a particular chemical element in the organic matter equalled by a corresponding decrease in the mineral soil has not been demonstrated. In fact, the amounts of exchangeable nutrients and total nitrogen in the mineral soil may be increased in high-producing woodlands having a large nutrient uptake. Long-term changes in woodland soils cannot be interpreted solely in terms of nutrient incorporation within the organic mass and loss from the soil, but account must be taken of the complete ecosystem dynamics, particularly the external flow of nutrients into and out of the woodland ecosystem. Any evaluation of the effect of woodlands on the soil resources must also embrace the complete crop rotation, from establishment to felling, since, when the tree trunks are harvested, the breakdown of the tree crowns, roots and litter left in the ecosystem would partly compensate for any depletion of the soil reserves that has occurred.

E. INPUT

Nutrients contained in the precipitation constitute an important addition to woodland ecosystems (Table XVIII) and, except for nitrogen and phosphorus, and to a lesser extent calcium, the amounts are not much less than the annual removal in tree trunks. When precipitation is intercepted by vegetation some of the contained nutrients may be absorbed directly by the vegetation. On balance, it seems that there is a greater leaching out of nutrients from the plants, since the amount contained in the precipitation is increased by interception, even though only a proportion of the rain-water reaches ground level (Madgwick and

Ovington, 1959). Whilst Tukey *et al.* (1958) have demonstrated leaching of nutrients from the foliage, part of the increase in the nutrient content of rain-water may be due to the washing off of aerosols and extraneous matter deposited on the vegetation. The large and complex woodland canopy forms an effective filter of the lower atmosphere which is becoming increasingly polluted with industrial wastes.

Tamm and Troedsson (1955) have drawn attention to the large amounts of material blown into woodland ecosystems from dirt roads resulting in an increase of the nutrient budget. Holstener-Jorgensen (1960) found that the annual input of wind-borne material into a spruce woodland in Denmark amounted to 1 000 kg per ha at a distance of 80 m from the woodland edge. He attributes this to soil blown from arable land and points out that, since the smaller, more nutrient-rich soil particles are carried by the wind, the addition of nutrients in this way must be considerable.

Additional supplies of nutrients are made available in the soil by weathering of soil particles and by soil formation from the underlying rock. The rate at which these two processes take place in woodlands is not known accurately but Klausing (1956) has suggested that in old beech woods, erosion of the parent rock may be as much as 1·2 mm a year for granite and 2·1 mm for diorite. The rate of weathering would be expected to be fairly high in woodland ecosystems because of the large annual turnover of organic matter and the huge root systems of trees which frequently extend deeply into rock fissures. In areas, where high winds are frequent, trees may be blown down (Lutz, 1940) and the resultant mass turnover of soil is of considerable significance in weathering, and in bringing the lower soil horizons to the surface. By this means compacted horizons may be shattered so that the effective soil volume and hence amount of nutrients available is increased. On steep slopes soil movement downslope may occur following windblow.

Since the gain of nitrogen by the forest from precipitation is small, most of the nitrogen must be obtained from the atmosphere. An indication of the magnitude of atmospheric nitrogen fixation by microorganisms is given by the fact that the annual accumulation of nitrogen in the organic material may be as much as 60 kg per ha and in addition there tends to be more nitrogen in the mineral soil after afforestation. Crocker and Major (1955) investigating a natural vegetation sequence, following glacier retreat, found that the average annual rate of nitrogen accumulation in the mineral soil under alder amounted to 26 kg per ha and comparable rates of accumulation have been recorded for forest plantations (Holmsgaard, 1960). Substantially more nitrogen must be fixed than is accumulated within the ecosystem. High rates of nitrogen fixation occur when forest vegetation contains plants, such as alder,

TABLE XVIII

Annual Supply of Chemical Elements in the Precipitation Falling on and below Forest Canopies

	Elements (kg per ha)						Country	Reference
	Na	K	Ca	Mg	P	N		
In precipitation	4–38	1–4	6–19	—	—	0·8–4·9	Sweden	Emanuelsson *et al.*, 1954
In precipitation	19	3	11	<4	<0·4	—	England	Madgwick and Ovington, 1959
Below hardwood canopies	31	28	25	11	—	—	England	Madgwick and Ovington, 1959
Below softwood canopies	34	23	24	9	—	—	England	Madgwick and Ovington, 1959
In precipitation	36	5	3	—	0·3	—	New Zealand	Will, 1959
Below *Pinus radiata* canopy	86	28	5	—	0·7	—	New Zealand	Will, 1959
In precipitation	40	10	4	—	0·6	—	New Zealand	Will, 1959
Below *Pseudotsuga taxifolia* canopy	40	30	8	—	6·9	—	New Zealand	Will, 1959
In precipitation	63	7	7	11	0·2	2·8	New Zealand	Miller, in manuscript
Below *Nothofagus truncata* canopy	74	31	13	13	0·6	2·3	New Zealand	Miller, in manuscript

bearing root nodules so that Bond (1958) has suggested that alder be used as a nurse crop on infertile land. However, nitrogen fixation may be considerable in pure stands of coniferous tree species, such as spruce, with no ground vegetation present. Tamm (1953) after studying the nutrition of the woodland moss *Hylocomium splendens* concludes that there may be a direct absorption of ammonia from the air and this may account for part of the increase in nitrogen of tree leaves placed on the forest floor by Gilbert and Bocock (1960).

On balance the nutrient capital of woodland ecosystems is probably increased by animal movement, particularly where the woodlands occur as small tracts within open country. Many birds roost in trees and their droppings are scattered over the vegetation and forest floor. Ailing animals may seek the shelter of woodlands where they die.

Appreciable amounts of nutrients may be added to forest ecosystems by man, either within seedlings transplanted to the forest from nurseries or by the direct application of fertilizers. Artificial fertilizers may be more generally applied in the future but at present their main use is to assist the establishment of young plants on infertile soils or to improve the growth of young woodlands exhibiting serious nutrient deficiency symptoms. Numerous examples of improved growth of seedlings or young trees following the application of fertilizers are given in a comprehensive bibliography compiled by White and Leaf (1956). Recently, increased attention has been paid to the possibilities of applying fertilizers to middle-aged and older, close-canopied woodlands in order to improve productivity and to reduce any soil degradation resulting from raising fast-growing woodlands on poor soils. Various techniques, including aerial spreading (Holmes and Cousins, 1960) have been developed to give an economic and even application of fertilizers over wooded areas.

The results obtained from applying nutrients to older woodlands vary greatly, in most cases an increase in nutrient uptake, indicated by foliar analysis, occurs, but this is not always accompanied by increased production. Leibundgut and Richard (1957) in Switzerland found no evidence of improved growth after applying various fertilizers to 45- to 70-year-old spruce growing on a rich, brown, forest soil. In Britain an elaborate series of experiments in which K, Ca, Mg, P and N were applied at various levels up to 250, 1 570, 35, 135 and 200 kg per ha respectively to pole stage crops of Sitka spruce, Scots pine and Douglas fir for a range of site conditions has failed to give any growth response by the trees during the first two years (personal communication from G. D. Holmes, Forestry Commission). In contrast, Heilman (1961) reports a marked effect of nitrogen supplied to natural Douglas fir stands on poor sites, the maximum application of 740 kg of N per ha increased tree growth 2 to 3 times over a 7- to 9-year period. The growth increase was associated with an increase in the weight of leaves carried by the forest and although the nitrogen content in the trees doubled that of phosphorus showed little change. Determination of the total nitrogen within the ecosystem suggested that little of the applied nitrogen was lost by leaching. Mayer-Krapoll (1956) gives the results of several German forest experiments showing improved tree growth following the application of nitrogen. Weston (1958) found that a top dressing of phosphate increased the growth of intermediate quality stands of *Pinus radiata* in New Zealand and Van Goor (1955) has stressed the need to consider the very different phosphorus demands of trees of different species when applying phosphate fertilizers. Positive growth responses by red pine to potassium have been reported by Heiberg *et al.* (1959) and to calcium by Hausser (1956). Liming, particularly when accompanied by nitrogen application, can have a very dramatic effect in increasing litter breakdown (Lohwasser, 1953) but there is a danger that over-liming may decrease tree growth (Van Goor, 1953).

The effect of increasing the supply of one element upon the availability and uptake of others has received scant attention and little is known of the long-term effects of fertilizer application particularly in relation to the dynamics of the ecosystem as a whole. It is evident that the effects of fertilizers are not confined to the trees, since different treatments stimulate or depress the production of fruit bodies by agarics (Hora, 1959) and change the ground flora (Malmström, 1943) and microfauna (Franz, 1957). More intensive and comprehensive research utilizing the ecosystem approach is needed to place the technique of fertilizer application on a more scientific basis, and to elucidate the interactions of different chemical elements.

F. OUTPUT

Previously, data have been given of the loss of nutrients by the cropping of tree boles but harvesting of foliage, game and litter may cause some drain on the nutrient capital. Another important factor on the debit side of the nutrient budget of woodlands is the removal of nutrients in the drainage water but unfortunately virtually no quantitative data are available. Since a forest cover reduces the outflow of water from catchment areas the total loss of nutrients in this way may be less under woodlands than under other types of plant cover but many diverse factors are concerned. On infertile soils there would apparently be greater competition between woodland plants for the limited amounts of nutrients available so that fortunately soil leaching may be at a minimum in these areas. Viro (1953) has calculated the loss of nutrients in Finland, which is largely forest covered, from data of the chemical composition and flow of river water. He found that run-off resulted in an annual loss of 4 kg per ha of K, 12 of Ca, 4 of Mg, 0·25 of P and 2 of N, in the case of the four mineral elements this exceeded input as precipitation by 2, 10, 3 and 0·17 kg per ha respectively, but twice as much nitrogen was supplied in the rain-water as was lost in the drainage water. Some nutrients in river water are derived from agricultural areas and from the underlying rock so that the loss of nutrient from forest ecosystems may be less than Viro's figures suggest. Miller (in manuscript) calculates that, in stands of *Nothofagus truncata* in New Zealand, leaching losses amount to about 30 kg per ha for Ca, 15 each for Mg and K and very little for phosphorus and nitrogen.

Forest fires may cause a large loss of nutrients from the ecosystem since volatile elements are released into the atmosphere and the mineral elements remaining in the ash may be washed out fairly rapidly (Burns, 1952) or dispersed by wind erosion. Under normal conditions the loss of nutrients through erosion tends to be negligible in wooded areas because of the blanketing effect of the vegetation and the litter. Serious erosion may occur at certain stages, e.g. at the time of harvesting because of soil disturbance due to the skidding out of logs or because of badly-located extraction roads.

G. Long-term Balance

It is impossible to generalize on long-term trends in the nutrient budget, since so many diverse processes are concerned in the circulation of chemical elements in woodland ecosystems, and all of these vary according to the character of the ecosystem, the particular element concerned and the influence of man. There may be no long-term balance of the nutrient capital even in natural woodlands. It does appear that the ecosystem capital for some essential nutrients, e.g. calcium, may be seriously depleted by intensive forestry so that high levels of production can only be maintained if sylviculturalists add nutrients to the system, as is done in agriculture. Much would be gained by bringing together the present fragmentary research on nutrient circulation in order to understand more fully the dynamic interrelationships of its many facets.

VII. Conclusions

One purpose of this essay is to stress the fundamental unity of ecosystem physiology, yet paradoxically, for convenience of description, the four processes described have had to be considered separately. No such distinction occurs in nature and many interrelationships exist, some of these were mentioned in passing, but the interconnections are so numerous and diverse that it has been impossible to describe them adequately. A change in any one process inevitably causes widespread repercussions in the rest.

The essentially dynamic nature of woodland ecosystems is masked by the impression of stability and permanence engendered by the longevity of trees. Yet woodland ecosystems are characterized by a mass seasonal transfer of organic matter, energy, water and chemical elements, the seasonal magnitudes of which change greatly as the trees mature and are replaced by trees of different species. There is a delicate and everchanging relationship between woodland organisms and their physical environment of soil and microclimate. Now that man is making fuller use of woodlands for cropping, hunting, recreation, etc. he is inducing long-term changes in woodland ecosystems and modifying the circulatory systems increasingly.

Despite rapid and significant advances during the last decade, no reasonably comprehensive and balanced picture of the quantitative ecology of a single woodland ecosystem exists. Nor is one likely to be produced in the near future because of the differences in scale, for instance of botanical, pedological or zoological studies, which create special problems of co-operation and integration. In the meantime, the task of synthesizing fragmentary and incomplete knowledge, gained by different disciplines, probably represents the most challenging and rewarding problem facing woodland ecologists. The ecosystem approach serves not only as a basis for evaluating and bringing together the existing diverse data but also acts as a guide for future research by spotlighting omissions and indicating the most fruitful fields for intensive study. Moreover, it draws the attention of scientists concerned with limited aspects of woodland ecology to the need to obtain and express their results in such a way that integration with other data is possible and meaningful.

The application of the ecosystem approach to woodland ecology is opportune. At the present time forestry is becoming more intensive and sylviculturalists are faced with the problem of producing not only greater amounts of cellulose per unit area of land but also of meeting the ever growing demand for more complete and multipurpose use of forest land. Sound principles of multipurpose use must be based on an improved appreciation of the complexity and dynamics of woodland ecosystems. It is significant that despite the lack of precise means of delimiting ecosystems, scientists concerned with the problem of classifying forest land for practical purposes are looking increasingly towards the ecosystem concept (Hustich, 1960) as a basis of classification. A better knowledge of woodland ecosystem dynamics is also needed because woodlands, by virtue of their size and mass transfer of material, possess certain unique attributes enabling them to serve as models for fundamental studies, such as those of Major (1951) and Jenny (1958), into the interplay of climate, soil and vegetation. The recent peaceful and martial developments of atomic energy with the mass release of radioactive material demands a better understanding of the flow of materials through and out of woodland ecosystems (Olson, 1959) if man is to survive. Although the problems to be solved in woodland ecology are so varied and immense, they present ecologists with both a challenge and a means of demonstrating the practical application of their subject for the welfare of man.

References

Alway, F. J. (1930). *J. For.* **28**, 715–727. Quantity and nutrient contents of pine leaf litter.

Alway, F. J. and Harmer, P. J. (1927). *Soil Sci.* **23**, 57–71. Minnesota glacial soil studies: II. The forest floor on the late Wisconsin Drift.

Alway, F. J. and Rost, C. O. (1927). *Soil Sci.* **3**, 546–576. Effect of forest fires upon the composition and productivity of the soil. Proc. First Int. Cong.

Armstrong, D. G. (1960). *Proc. Eighth Int. Grassland Congr.*, 485–489. Calorimetric determination of the net energy value of dried S.23 ryegrass at four stages of growth.

Aubréville, A. (1938). *Ann. Acad. Sci. colon., Paris* **9**, 1–245. La forêt colonial: les forêts de L'Afrique occidentale française.

Auerbach, S. I. (1958). *Ecology* **39**, 522–529. The soil ecosystem and radioactive waste disposal to the ground.

Bates, C. G. and Henry, A. J. (1928). *Mon. Weath. Rev., Wash. Suppl.* **30**, 1–79. Forest and stream-flow at Wagon Wheel Gap, Colorado.

Billings, W. D. and Bliss, L. C. (1959). *Ecology* **40**, 388–397. An alpine snowbank environment and its effects on vegetation, plant development, and productivity.

Blackman, G. E. and Rutter, A. J. (1947). *Ann. Bot., Lond.* **N.S. 11**, 125–158. Physiological and ecological studies in the analysis of plant environment.

Bleasdale, A. (1957). *Emp. For. Rev.* **36**, 59–66. Afforestation of catchment areas; the Physicists approach to problems of water loss from vegetation.

Bocock, K. L. and Gilbert, O. J. W. (1957). *Plant and Soil* **9**, 179–185. The disappearance of leaf litter under different woodland conditions.

Bocock, K. L., Gilbert, O. J. W., Capstick, C. K., Twinn, D. C., Waid, J. S. and Woodman, M. S. (1960). *J. Soil Sci.* **11**, 1–9. Changes in leaf litter when placed on the surface of soils with contrasting humus types.

Bond, G. (1958). Symbiotic nitrogen fixation by non-legumes, 216–231. Reprint from "Nutrition of the Legumes" (E. G. Hallsworth, ed.). Butterworth Scientific Publications.

Bormann, F. H. and Graham Jr., B. F. (1959). *Ecology* **40**, 677–691. The occurrence of natural root grafting in eastern white pine, *Pinus strobus* L., and its ecological implications.

Bornebusch, C. H. (1930). *Forstl. Forsøksv. Danm.* **11**, 1–224. The fauna of forest soil.

Bourdeau, P. F. (1959). *Ecology* **40**, 63–67. Seasonal variations of the photosynethetic efficiency of evergreen conifers.

Bray, J. R., Lawrence, D. B. and Pearson, L. C. (1959). *Oikos* **10**, 38–49. Primary production in some Minnesota terrestrial communities.

Brown, R. T. and Curtis, J. T. (1952). *Ecol. Monogr.* **22**, 217–234. The upland conifer-hardwood forests of northern Wisconsin.

Burger, H. (1929). *Mitt. schweiz. ZentAnst. forstl. Versuchsw.* **15**, 243–292. Holz, Blattmenge und Zuwachs. 1. Die Weymouthsföhre.

Burger, H. (1935). *Mitt. schweiz. ZentAnst. forstl. Versuchsw.* **19**, 21–72. Holz, Blattmenge und Zuwachs. 2. Die Douglasie.

Burger, H. (1945). *Mitt. schweiz. ZentAnst. forstl. Versuchsw.* **24**, 7–103. Holz, Blattmenge und Zuwachs. 7. Die Lärche.

Burger, H. (1947). *Mitt. schweiz. ZentAnst. forstl. Versuchsw.* **25**, 211–279. Holz, Blattmenge und Zuwachs. 8. Die Eiche.

Burger, H. (1948). *Mitt. schweiz. ZentAnst. forstl. Versuchsw.* **25**, 435–493. Holz, Blattmenge und Zuwachs. 9. Die Föhre.

Burger, H. (1950). *Mitt. schweiz. ZentAnst. forstl. Versuchsw.* **26**, 419–497. Holz, Blattmenge und Zuwachs. 10. Die Buche.

Burger, H. (1951). *Mitt. schweiz. ZentAnst. forstl. Versuchsw.* **27**, 247–286. Holz, Blattmenge und Zuwachs. 11. Die Tanne.

Burger, H. (1953). *Mitt. schweiz. ZentAnst. forstl. Versuchsw.* **29**, 38–128. Holz, Blattmenge und Zuwachs. 13. Fichten im gleichalterigen Hochwald.

Burger, H. (1954). *Mitt. schweiz. ZentAnst. forstl. Versuchsw.* **31**, 9–58. Einfluss des Waldes auf den Stand der Gewässer.

Burges, A. (1950). *Trans. Brit. mycol. Soc.* **33**, 142–147. The downward movement of fungal spores in sandy soil.

Burns, P. Y. (1952). *Yale Sch. For. Bull.* **57**, 50 pp. Effect of fire on forest soils in the pine-barren regions of New Jersey.

Christie, J. M. and Lewis, R. E. A. (1961). *For. Rec.* **47**, 48 pp. Provisional yield tables for *Abies grandis* and *Abies nobilis*.

Crocker, R. L. (1952). *Quart. Rev. Biol.* **27**, 139–168. Soil genesis and the pedogenic factors.

Crocker, R. L. and Major, J. (1955). *J. Ecol.* **43**, 427–448. Soil development in relation to vegetation and surface age at Glacier Bay, Alaska.

Croft, A. R. and Monninger, L. V. (1953). *Trans. Amer. geophys. Un.* **34**, 563–574. Evapotranspiration and other water losses on some aspen forest types in relation to water available for stream flow.

Crossley Jr., D. A. and Howden, H. F. (1961). *Ecology* **42**, 302–317. Insect vegetation relationships in an area contaminated by radio-active wastes.

Curtis, J. T. and McIntosh, R. P. (1951). *Ecology* **32**, 476–496. An upland forest continuum in the prairie-forest border region of Wisconsin.

Dadykin, V. P. (1960). "Utilisation of light by tree species depending on the external conditions", 72–77. Questions of forestry and forest management. U.S.S.R. Academy of Sciences, Moscow.

Davis, K. P. (1959). "Forest Fire, Control and Use", 584 pp. McGraw Hill.

Day, G. M. (1940). *J. For.* **38**, 646–648. Topsoil changes in coniferous plantations.

de Wit, C. T. (1959). *Ned. J. Agr. Sci.* **7**, 141–149. Potential photosynthesis of crop surfaces.

Dimbleby, G. W. (1952). *J. Ecol.* **40**, 331–341. Soil regeneration on the north-east Yorkshire moors.

Dimbleby, G. W. (1961). *J. Soil Sci.* **12**, 1–11. Soil pollen analysis.

Dreibelbis, F. R. and Post, F. A. (1941). *Proc. Soil Sci. Soc. Amer.* **6**, 462–473. An inventory of soil water relationships on woodland, pasture and cultivated soils.

Duchaufour, Ph., and Guinaudeau, J. (1957). *Ann. Éc. Eaux For. Nancy* **15**, 336–364. Une expérience de chaulage sur humus brût.

Dunford, E. G. (1954). *J. For.* **52**, 923–927. Surface runoff and erosion from pine grasslands of the Colorado Front Range.

Ehwald, E. (1957). Über den Nährstoffkreislauf des Waldes. *Dtsch. Akad. Landwirtschaftswissenschaften Berlin* **6**, 1–56.

Ellison, L. and Houston, W. P. (1958). *Ecology* **39**, 337–345. Production of herbaceous vegetation in openings and under canopies of western aspen.

Emmanuelsson, A., Eriksson, E. and Egnér, H. (1954). *Tellus* **3**, 261–267. Composition of atmospheric precipitation in Sweden.

Evans, F. C. (1956). *Science* **123** (*3208*), 1127–1128. Ecosystem as the basic unit in ecology.

Franz, H. (1957). *Allg. Forstz.* **68**, 58–63. Present knowledge on the use of fertilizers in forestry.

Fraser, D. A. (1952). *Pulp Pap.* (*Mag.*) *Can.* 202–209. Growth mechanisms in hardwoods.

Fritts, H. C. (1958). *Ecology* **39**, 705–720. An analysis of radial growth of beech in a central Ohio forest during 1954–1955.

Gibb, J. A. (1960). *Ibis* **102**, 163–208. Populations of tits and goldcrests and their food supply in pine plantations.

Gilbert, O. J. W. and Bocock, K. L. (1960). *J. Soil Sci.* **11**, 10–19. Changes in leaf litter when placed on the surface of soils with contrasting humus types. 2. Changes in the nitrogen content of oak and ash leaf litter.

Giulimondi, G., Funiciello, M. and Arru, G. M. (1956). *Pubbl. Cen. Sper. Agric. For.* **1**, 111–130. Richerche sui terreni coltivati and eucalitti.

Golley, F. B. (1960). *Ecol. Monogr.* **30**, 187–206. Energy dynamics of a food chain of an old field community.

Golley, F. B. (1961). *Ecology* **42**, 581–584. Energy values of ecological materials.

Greenland, D. J. and Kowal, J. M. L. (1960). *Plant & Soil* **12**, 154–174. Nutrient content of the moist tropical forest of Ghana.

Grimshaw, H. M., Ovington, J. D., Betts, M. M. and Gibb, J. A. (1958). *Oikos* **9**, 26–34. The mineral content of birds and insects in plantations of *Pinus sylvestris* L.

Haberland, F. P. and Wilde, S. A. (1961). *Ecology* **42**, 584–586. Influence of thinning of red pine plantations on soil.

Härtel, O. and Rudolf, E. (1953). *Zbl. ges. Forst- u. Holzw.* **72**, 47–59. Zur Physiologie und Ökologie der Kutikulären Wasseraufnahme durch Koniferennaden.

Harley, J. L. (1959). "The Biology of Mycorrhiza", 233 pp. Plant Science Monograph. Leonard Hill Books Ltd.

Hausser, K. (1956). *Phosphorsäure* **16**, 9–27. Ertragsteigerung in der Forstwirtschaft durch mineralische Düngung.

Heiberg, S. O., Leyton, L. and Loewenstein, H. (1959). *For. Sci.* **5**, 142–153. Influence of potassium fertiliser level on red pine planted at various spacings on a potassium-deficient site.

Heilman, P. E. (1961). Effects of nitrogen fertilization on the growth and nitrogen nutrition of low-site Douglas Fir stands. Ph.D. Thesis, University of Washington.

Hellmers, H. and Bonner, J. (1959). *Proc. Soc. Amer. For.*, 32–35. Photosynthetic limits of forest tree yields.

Hiley, W. E. (1954). "Woodland Management", 463 pp. Faber and Faber, London.

Hills, G. A. (1960). "The Total Site Classification of Forest Productivity", 61 pp. Research Branch Paper. Ontario Dept. of Lands and Forests.

Höfler, K. (1937). *Ber. dtsch. bot. Ges.* **55**, 606–622. Pilsoziologie.

Holmes, G. D. and Cousins, D. A. (1960). *Forestry* **33**, 54–73. Application of fertilizers to checked plantations.

Holmsgaard, E. (1960). *Forstl. Forsøksv. Danm.* **26**, 253–270. Kvaelstofbindingens størrelse hos el litteraturgennemgang og en undergelse af et plantningsforsøg.

Holstener-Jørgensen, H. (1960). *Forstl. Forsøksv. Danm.* **26**, 391–397. Indfygning af jord i en plantages vestrand.

Hoover, M. D. (1944). *Trans. Amer. Geophys. Un.* **25**, 969–975. Effect of removal of forest vegetation upon water yields.

Hoover, M. D. (1953). Interception of rainfall in a young Loblolly pine plantation, 13 pp. U.S. D.A. For. Ser. Southeast. For. Expt. Sta. Pap. 21.

Hopkins, B. (1960). *E. Afr. Agric. J.* **25**, 255–258. Rainfall interception by a tropical forest in Uganda.

Hora, F. B. (1959). *Trans. Brit. mycol. Soc.* **42**, 1–14. Quantitative experiments on toadstool production in woods.

Huber, B. (1950). *Ber. dtsch. bot. Ges.* **63**, 53–64. Registrierung des CO_2-Gefälles und Berechnung des CO_2-Stromes über Pflanzengesellschaften mittels Ultrarot Absorptionsschreiber.

Hustich, I. (1960). *Silva fenn.* **105**. Forest types and forest ecosystems. A symposium of the IX International Botanical Congress, 142 pp.

Ivanov, L. A., Silina, A. A., Zhmur, D. G. and Tselniker, Y. U. (1951). *Bot. Zh. S.S.S.R.* **36**, 5–20. Determination of transpiration loss by a forest stand.

Jenny, H. (1958). *Ecology* **39**, 5–16. Role of the plant factor in the pedogenic functions.

Johansson, N. (1933). *Svenska SkogsvFören Tidskr.* **31**, 53–134. The relation between the respiration of the tree stem and its growth.

Johnson, E. A. and Kovner, J. L. (1956). *For. Sci.* **2**, 82–91. Effect on stream-flow of cutting a forest understorey.

Jones, E. W. (1956). *J. Ecol.* **43**, 564–594. Ecological studies on the rain forest of Southern Nigeria. IV. The plateau forest of the Okomu Forest Reserve.

Klausing, O. (1956). *Forstwiss. Zbl.* **75**, 18–32. Untersuchungen über den Mineralumsatz in Buchenwäldern auf Granit und Diorit.

Kozlowski, T. T. and Ward, R. C. (1957a). *For. Sci.* **3**, 61–66. Seasonal height growth of conifers.

Kozlowski, T. T. and Ward, R. C. (1957b). *For. Sci.* **3**, 168–174. Seasonal height growth of deciduous trees.

Kramer, P. J. (1958). Photosynthesis of trees as affected by their environment. The physiology of forest trees. 157–181. "Harvard Forest Symposium" (K. V. Thimann, ed.). Ronald Press Company.

Lack, D. (1933). *J. Anim. Ecol.* **2**, 239–262. Habitat selection in birds, with special reference to the effects of afforestation on the Breckland avifauna.

Lack, D. (1939). *J. Anim. Ecol.* **8**, 277–285. Further changes in the Breckland avifauna caused by afforestation.

Ladefoged, K. (1956). *Dansk. Skovforen. Tidsskr.* **41**, 481–506. Studies on water consumption by trees.

Larsen, C. S. (1956). "Genetics in Silviculture", 224 pp. Translated by M. L. Anderson. Oliver and Boyd.

Law, F. (1956). *J. Brit. Waterwks. Ass.* **38**, 489–494. The effect of afforestation upon the yield of water catchment areas.

Leibundgut, H. and Richard, F. (1957). *Schweiz. Z. Forstw.* **108**, 129–144. Beitrag zum Problem der Düngung im schweizerischen Waldbau.

Lindeman, R. L. (1942). *Ecology* **23**, 399–418. The trophic-dynamic aspect of ecology.

Lohwasser, W. (1953). *Forstarchiv* **24**, 213–222. Kalkdüngsversuche im Eggegbirge und Hunsrück.

Lounamaa, J. (1956). *Ann. Bot. Soc. Vanamo* **29**, 1–196. Trace elements in plants growing wild on different rocks in Finland.

Lull, H. W. (1959). *J. For.* **57**, 905–909. Humus depth in the north-east.

Lutz, H. J. (1940). *Yale Sch. For. Bull.* **45**, 37 pp. Disturbance of forest soil resulting from the uprooting of trees.

Lutz, H. J. and Chandler, R. F. (1946). Forest soils. 514 pp. J. Wiley & Sons.

Macfadyen, A. (1957). "Animal Ecology", 264 pp. Pitman.

Macfadyen, A. (1961). *Ann. appl. Biol.* **49**, 215–218. Metabolism of soil invertebrates in relation to soil fertility.

Madgwick, H. A. I. and Ovington, J. D. (1959). *Forestry* **32**, 14–22. The chemical composition of precipitation in adjacent forest and open plots.

Major, J. (1951). *Ecology* **32**, 392–414. A functional, factorial approach to plant ecology.

Malmström, C. (1943). *Norrlands SkogsvForb. Tidskr.* **4**, 273–292. Skogliga gödslingsförsök pa dikade svaga torvmarker.

Mayer-Krapoli. H. (1956). "The Use of Commercial Fertilisers — particularly Nitrogen — in Forestry", 111 pp. Allied Chemical and Dye Corp., New York.

Metz, L. J. (1954). *Soil Sci. Soc. Amer. Proc.* **18**, 335–338. Forest floor in the Piedmont Region of South Carolina.

Miller, R. B. and Hurst, F. B. (1957). *N.Z. For. Res. Notes* **8**, 14 pp. The quantity and nutrient content of hard beech litter.

Mina, V. N. (1955). *Pochvovedenie* **6**, 32–44. Cycle of N and ash elements in mixed oakwoods of the forest-steppe.

Möller, C. M. (1947). *J. For.* **45**, 393–404. The effect of thinning, age and site on foliage, increment and loss of dry matter.

Möller, C. M. (1954). *Coll. For. Bull. Syracuse* **1**, 5–44. The influence of thinning and volume increment.

Möller, C. M. (1960). *Forestry* **33**, 37–53. The influence of pruning on the growth of conifers.

Möller, C. M., Müller, D. and Nielsen, J. (1954). *Forstl. Forsøksv. Danm.* **21**, 273–301. Respiration in stem and branches of beech.

Monteith, J. L. (1959). *Quart. J. R. met. Soc.* **85**, 386–392. The reflection of short-wave radiation by vegetation.

Ničiporovič, A. A. (1960). *Field Crop Abstr.* **13**, 169–175. Photosynthesis and the theory of obtaining high crop yields. An abstract with commentary by J. N. Black and D. J. Watson.

Ničhiporovičh, A. A. and Strogonova, L. E. (1957). *Agrochimica* **2**, 26–53. Photosynthesis and problems of crop yield.

Nielsen, C. O. (1955). Studies on Enchytraeidae. 2. Field Studies, 58 pp. Naturhistorisk Museum, Aarhus.

Oberlander, G. T. (1956). *Ecology* **37**, 851–852. Summer fog precipitation of the San Francisco peninsula.

O'Connor, F. B. (1957). *Oikos* **8**, 161–197. An ecological study of the Enchytraeid worm population of a coniferous forest soil.

Odum, E. P. (1960). *Ecology* **41**, 34–49. Organic production and turnover in old field succession.

Odum, H. T. (1957). *Ecol. Monogr.* **27**, 55–112. Trophic structure and productivity of Silver Springs, Florida.

Odum, H. T. and Odum, E. P. (1955). *Ecol. Monogr.* **25**, 291–320. Trophic structure and productivity of a windward coral reef community on Eniwetok Atoll.

Ogawa, H., Yoda, K. and Kira, T. (1961). *Nature and Life in South east Asia* **1**, 21–157. A preliminary survey on the vegetation of Thailand.

Olson, J. S. (1959). Environmental Radioactivity. Progress in Ecological Research related to radioactive waste disposal and fall-out. Forest Studies. Health Physics Division Annual Progress Report. ORNL-2806. U.S. Atomic Energy Commission.

Orlov, A. Ja. (1955). *Pochvovedenie* **6**, 14–20. The role of feeding roots of forest vegetation in enriching soils with organic matter.

Ovington, J. D. (1954a). *J. Ecol.* **42**, 71–80. Studies of the development of woodland conditions under different trees. 2. The forest floor.

Ovington, J. D. (1954b). *Forestry* **27**, 41–53. A comparison of rainfall in different woodlands.

Ovington, J. D. (1955). *J. Ecol.* **43**, 1–21. Studies of the development of woodland conditions under different trees. 3. The ground flora.

Ovington, J. D. (1956a). *New Phytol.* **55**, 289–304. The form, weights and productivity of tree species grown in close stands.

Ovington, J. D. (1956b). *J. Ecol.* **44**, 171–179. Studies of the development of woodland conditions under different trees. 4. The ignition loss, water, carbon and nitrogen contents of the mineral soil.

Ovington, J. D. (1956c). *J. Ecol.* **44**, 597–604. Studies of the development of woodland conditions under different trees. 5. The mineral composition of the ground flora.

Ovington, J. D. (1957a). *Ann. Bot., Lond.* **N.S. 21**, 287–314. Dry matter production by *Pinus sylvestris* L.

Ovington, J. D. (1957b). *New Phytol.* **56**, 1–11. The volatile matter, organic carbon and nitrogen contents of tree species grown in close stands.

Ovington, J. D. (1958a). *New Phytol.* **57**, 273–284. The sodium, potassium and phosphorus contents of tree species grown in close stands.

Ovington, J. D. (1958b). *J. Ecol.* **46**, 127–142. Studies of the development of woodland conditions under different trees. 6. Soil sodium, potassium and phosphorus.

Ovington, J. D. (1958c). *J. Ecol.* **46**, 391–405. Studies of the development of woodland conditions under different trees. 7. Soil calcium and magnesium.

Ovington, J. D. (1959a). *Ann. Bot., Lond.* **N.S. 23**, 229–239. The circulation of minerals in plantations of *Pinus sylvestris* L.

Ovington, J. D. (1959b). *Ann. Bot., Lond.* **N.S. 23**, 75–88. Mineral content of plantations of *Pinus sylvestris* L.

Ovington, J. D. (1959c). *New Phytol.* **58**, 164–175. The calcium and magnesium contents of tree species grown in close stands.

Ovington, J. D. (1961). *Ann. Bot., Lond.* **N.S. 25**, 12–20. Some aspects of energy flow in plantations of *Pinus sylvestris* L.

Ovington, J. D. and Heitkamp, D. (1960). *J. Ecol.* **48**, 639–646. The accumulation of energy in forest plantations in Britain.

Ovington, J. D. and MacRae, C. (1960). *J. Ecol.* **48**, 549–555. The growth of seedlings of *Quercus petraea*.

Ovington, J. D. and Madgwick, H. A. I. (1959a). *For. Sci.* **5**, 344–355. Distribution of organic matter and plant nutrients in a plantation of Scots pine.

Ovington, J. D. and Madgwick, H. A. I. (1959b). *Plant & Soil* **10**, 271–283. The growth and composition of natural stands of birch. 1. Dry matter production.

Ovington, J. D. and Madgwick, H. A. I. (1959c). *Plant & Soil* **10**, 389–400. The growth and composition of natural stands of birch. 2. The uptake of mineral nutrients.

Owen, H. T. (1954). *Forestry* **27**, 7–15. Observations on the monthly litter fall and nutrient content of sitka spruce litter.

Parker, J. (1961). *Ecology* **42**, 372–380. Seasonal trends in carbon dioxide absorption, cold resistance, and transpiration of some evergreens.

Pase, C. and Hurd, R. M. (1957). *Proc. Soc. Amer. For.*, 156–158. Understorey vegetation as related to basal area, crown cover and litter produced by immature Ponderosa Pine stands in the Black Hills.

Peace, T. R. (1961). *Advanc. Sci., Lond.* **17** (*69*), 448–455. The dangerous concept of the natural forest.

Pearson, L. C. and Lawrence, D. B. (1958). *Amer. J. Bot.* **45**, 383–387. Photosynthesis in aspen bark.

Penman, H. L. (1956). *Trans. Amer. geophys. Un.* **37**, 43–50. Estimating evaporation.

Pogrebnyak, P. S. (1960). Creation of mixed stands as a method for raising the productivity of forests, 104–112. Questions of forestry and forest management. U.S.S.R. Academy of Sciences, Moscow.

Polster, H. (1950). "Die physiologischen Grundlagen der Stofferzeugung im Walde", 96 pp. Bayerischer Landwirtschaftsverlag, Munich.

P'Yavchenko, N. I. (1960). *Pochvovedenie* **6**, 21–32. The biological cycle of nitrogen and mineral substances in bog forests.

Remezov, N. P. (1959). *Pochvovedenie* **1**, 71–79. Methods of studying the biological cycle of elements in the forest.

Rennie, P. J. (1955). *Plant & Soil* **7**, 49–95. Uptake of nutrients by mature forest growth.

Richards, P. W. (1952). "The Tropical Rain Forest", 450 pp. Cambridge University Press.

Rider, N. E. (1957). *Quart. J. R. met. Soc.* **83**, 181–193. Water losses from various land surfaces.

Rowe, J. S. (1961). *Ecology* **42**, 420–427. The level of integration concept and ecology.

Rowe, J. S., Haddock, P. G., Hills, G. A., Krajina, V. J. and Linteau, A. (1960). "The Ecosystem Concept in Forestry", 5 pp. Special paper. Fifth World Forestry Congress.

Russell, E. J. (1956). "Soil Conditions and Plant Growth", 635 pp. Longmans.

Rutter, A. J. (1957). *Ann. Bot., Lond.* **N.S. 21**, 399–426. Studies in the growth of young plants of *Pinus sylvestris* L. I. The annual cycle of assimilation and growth.

Rutter, A. J. (1959). *Int. Un. Geod. Geoph.* **48**, 101–110. Evaporation from a plantation of *Pinus sylvestris* in relation to meteorological and soil conditions.

Saeki, T. and Nomoto, T. (1958). *Bot. Mag., Tokyo* **71**, 235–241. On the seasonal change of photosynthetic activity of some deciduous and evergreen broadleaf trees.

Satoo, T. and Senda, M. (1958). *Bull. Tokyo Univ. For.* **54**, 71–100. Materials for the studies of growth in stands. IV. Amount of leaves and production of wood in a young plantation of *Chamaecyparis obtusa*.

Satoo, T., Nakamura, K. and Senda, M. (1955). *Bull. Tokyo Univ. For.* **48**, 63–90. Materials for the studies of growth in stands. I. Young stands of Japanese red pine of various density.

Satoo, T., Kunugi, R. and Kumekawa, A. (1956). *Bull. Tokyo Univ. For.* **52**, 33–58. Materials for the studies of growth in stands. 3. Amount of leaves and production of wood in an aspen (*Populus davidiana*) second growth in Hokkaido.

Satoo, T., Negesi, K. and Senda, M. (1959). *Bull. Tokyo Univ. For.* **55**, 101–123. Materials for the studies of growth in stands. V. Amount of leaves and growth in plantations of *Zelkowa serrata* applied with crown thinning.

Schwappach, A. (1912). "Ertragstafeln der wichtigeren Holzarten", 83 pp. Neudamn.

Scott, D. R. M. (1955). *Yale Sch. For. Bull.* **62**, 73 pp. Amount and chemical composition of the organic matter contributed by overstorey and understorey vegetation to forest soil.

Senda, M. and Satoo, T. (1956). *Bull. Tokyo Univ. For.* **52**, 15–31. Materials for the study of growth in stands. 2. White Pine (*Pinus strobus*). Stands of various densities in Hokkaido.

Shanks, R. E. and Olson, J. S. (1961). *Science* **134**, 194–195. First year breakdown of leaf litter in Southern Appalachian Forests.

Sjörs, H. (1954). *Acta. phytogeogr. suec.* **34**, 1–134. Slatterängar i Grangärde Finnmark.

Sjörs, H. (1955). *Svensk. bot. Tidskr.* **49**, 155–169. Remarks on ecosystems.

Smirnova, K. M. and Gorodentseva, G. A. (1958). *Bull. Soc. Nat. Moscou (biol.)* **62**, 135–147. The consumption and rotation of nutritive elements in birch woods.

Sonn, S. W. (1960). "Der Einfluss des Waldes auf die Böden", 166 pp. Gustav Fischer Verlag, Jena.

Specht, R. L., Rayson, P. and Jackson, M. E. (1958). *Aust. J. Bot.* **6**, 59–88. Dark Island Heath (Ninety-Mile Plain, South Australia). VI. Pyric succession: changes in composition, coverage, dry weight, and mineral nutrient status.

Stålfelt, M. G. (1944). *Särtryck ur Kungl. Lantbruksakademiens Tidskrift.* **83**, 1–83. Granens vattenförbrukning och dess inverkan på vattenomsattningen i marken.

Sukachev, V. N. (1944). *J. gen. Biol., Moscow* **5**, 213–277. On the principles of genetical classification in biocoenology.

Sukachev, V. N. (1960). Forest Biogeocenology as a theoretical basis for silviculture and forestry, 41–51. Questions of forestry and forest management. U.S.S.R. Academy of Sciences, Moscow.

Sutcliffe, R. C. (1956). *Quart. J. R. met. Soc.* **82**, 385–395. Water balance and the general circulation of the atmosphere.

Tadaki, Y. and Shidei, T. (1960). *J. Jap. For. Soc.* **42**, 427–434. Studies on productive structure of forest. 1. The seasonal variation of leaf amount and the dry matter production of deciduous sapling stand (*Ulmus parviflora*).

Tamm, C. O. (1953). *Medd. Skogsförsökanst. Stockh.* **43**, 1–140. Growth, yield and nutrition in carpets of a forest moss (*Hylocomium splendens*).

Tamm, C. O. and Carbonnier, C. (1961). *Skogs. och Lantbruksakademiens sammanträde tidskrift.* **100**, 95–124. Växtnäringen som skoglig produktionsfaktor. Kungl.

Tamm, C. O. and Östlund, H. G. (1960). *Nature, Lond.* **185**, 706–707. Radiocarbon dating of soil humus.

Tamm, C. O. and Troedsson, T. (1955). *Oikos* **6**, 61–70. An example of the amounts of plant nutrients supplied to the ground in road dust.

Tansley, A. G. (1935). *Ecology* **16**, 284–307. The use and abuse of vegetational concepts and terms.

Tranquillini, W. (1959a). *Planta* **54**, 107–129. Die Stoffproduktion der Zirbe (*Pinus cembra* L.) an der Waldgrenze während eines Jahres. I. Standortsklima und Co_2-assimilation.

Tranquillini, W. (1959b). *Planta* **54**, 130–151. Die Stoffproduktion der Zirbe (*Pinus cembra* L.) an der Waldgrenze während eines Jahres. II. Zuwaches und CO_2-bilanz.

Transeau, E. N. (1926). *Ohio J. Sci.* **26**, 1–10. The accumulation of energy by plants.

Tukey Jr., H. B., Tukey, H. B. and Wittwer, S. H. (1958). *Proc. Amer. Soc. hort. Sci.* **71**, 496–506. Loss of nutrients by foliar leaching as determined by Radioisotopes.

Turček, F. J. (1956). *Ibis* **98**, 24–33. On the bird population of the spruce forest community in Slovakia.

Van Goor, G. P. (1953). *Ned. BoschbTijdschr.* **25**, 57–68. Groeiremmingen bij de Japanse lariks (*Larix leptolepis*) ten gevolge van Kaklbemestingen.

Van Goor, C. P. (1955). *Het Thomasmeel* **11**, 251–257. De fosfaatbehoefte van bomen en de fosfaatbemesting in de bosbouw.

Vinokurov, M. A. and Tyurmenko, A. N. (1958). *Pochvovedenie* **7**, 787–791. Materials in the forest's biological cycle of nitrogen and phosphorus.

Viro, P. J. (1953). *Comm. inst. forest. Fenn.* **42**, 1–50. Loss of nutrients and the natural nutrient balance of the soil in Finland.

Voigt, G. K. (1960). *For. Sci.* **6**, 2–10. Distribution of rainfall under forest stands.

Wassink, E. C. (1959). *Plant Physiol.* **34**, 356–361. Efficiency of light energy conversion in plant growth.

Waters, W. T. and Christie, J. M. (1958). *For. Rec.* **36**, 31 pp. Provisional yield tables for oak and beech in Great Britain.

Watson, D. J. (1958). *Symp. Inst. Biol.* **7**, 25–32. Factors limiting production. The biological productivity of Britain.

Weston, G. C. (1958). *Proc. N.Z. Soc. Soil Sci.* **3**, 13–19. The response of Radiata pine to fertilizers.

White, D. P. and Leaf, A. L. (1956). Forest fertilization, 305 pp. *Technical Publication 81.* State University College of Forestry, Syracuse, New York.

Whittaker, R. H. (1953). *Ecol. Monogr.* **23**, 41–78. A consideration of climax theory: the climax as a population and pattern.

Whittaker, R. H. (1956). *Ecol. Monogr.* **26**, 1–80. Vegetation of the Smoky Mountains.

Whittaker, R. H. (1961). *Ecology* **42**, 177–180. Estimation of net primary production of forest and shrub communities.

Will, G. M. (1959). *N.Z. J. Agr. Res.* **2**, 719–734. Nutrient return in litter and rainfall under some exotic conifer stands in New Zealand.

Wind, R. (1958). *Committee on Hydrological Research T.N.O.* **3**, 164–228. The lysimeters in the Netherlands.

Winston, P. W. (1956). *Ecology* **37**, 120–32. The acorn microsere with special reference to arthropods.

Witherspoon, J. P. Unpublished work submitted for a Ph.D. degree at the Botany Department of the University of Tennessee.

Wittich, W. (1961). *Allg. Forstz.* **16**, 4 pp. Der Einfluss der Baumart auf den Bodenzustand.

Wright, T. W. and Will, G. M. (1958). *Forestry* **31**, 13–25. The nutrient content of Scots and Corsican Pines growing on sand dunes.

Zahner, R. (1955). *For. Sci.* **1**, 258–264. Soil water depletion by pine and hardwood stands during a dry season.

Zinecker, E. (1950). *Allg. Forst- u Jagdztg.* **5**, 139–144. The dynamic interpretation of forest pedology.

Zinke, P. J. (1959). *Ass. Int. Hydrol. Sci.* **49**, 126–138. The influence of a stand of *Pinus coulteri* on the soil moisture regime of a large San Dimas lysimeter in southern California.

Author Index

Numbers in italics refer to the pages on which references are listed in bibliographies at the end of each article.

Subject Index